Industrial Motor Control: Workbook and Lab Manual, 6E

Industrial Motor Control: Workbook and Lab Manual, 6E

Stephen L. Herman

Australia • Brazil • Japan • Korea • Mexico • Singapore • Spain • United Kingdom • United States

Industrial Motor Control Workbook and Lab Manual, 6th Edition
Stephen L. Herman

Vice President, Career and Professional Editorial: Dave Garza
Director of Learning Solutions: Sandy Clark
Managing Editor: Larry Main
Acquisitions Editor: Stacy Masucci
Senior Product Manager: John Fisher
Senior Editorial Assistant: Dawn Daugherty
Vice President, Career and Professional Marketing: Jennifer McAvey
Marketing Director: Deborah S. Yarnell
Marketing Manager: Jimmy Stephens
Marketing Coordinator: Mark Pierro
Production Director: Wendy Troeger
Production Manager: Mark Bernard
Content Project Manager: Christopher Chien
Art Director: Bethany Casey
Technology Project Manager: Christopher Catalina
Production Technology Analyst: Thomas Stover

Library of Congress Control Number: 2008935164

ISBN-13: 978-1-4354-4240-5
ISBN-10: 1-4354-4240-7

Delmar
5 Maxwell Drive
Clifton Park, NY 12065-2919
USA

Cengage Learning is a leading provider of customized learning solutions with office locations around the globe, including Singapore, the United Kingdom, Australia, Mexico, Brazil and Japan. Locate your local office at: **international.cengage.com/region**

Cengage Learning products are represented in Canada by Nelson Education, Ltd.

For your lifelong learning solutions, visit **delmar.cengage.com**
Visit our corporate website at **cengage.com**.

NOTICE TO THE READER
Publisher does not warrant or guarantee any of the products described herein or perform any independent analysis in connection with any of the product information contained herein. Publisher does not assume, and expressly disclaims, any obligation to obtain and include information other than that provided to it by the manufacturer. The reader is expressly warned to consider and adopt all safety precautions that might be indicated by the activities described herein and to avoid all potential hazards. By following the instructions contained herein, the reader willingly assumes all risks in connection with such instructions. The publisher makes no representations or warranties of any kind, including but not limited to, the warranties of fitness for particular purpose or merchantability, nor are any such representations implied with respect to the material set forth herein, and the publisher takes no responsibility with respect to such material. The publisher shall not be liable for any special, consequential, or exemplary damages resulting, in whole or part, from the readers' use of, or reliance upon, this material.

Printed in the United States of America
2 3 4 5 xx 12 11 10

TABLE OF CONTENTS

PREFACE

The *Student Workbook and Laboratory Manual for Industrial Motor Controls* is intended to give students the hands-on experience that is so vital in learning to draw, analyze, and connect motor control schematics. The worksheets and experiments are divided into separate sections. This permits the instructor to choose between worksheets and laboratory exercises. The worksheets are new for the sixth edition of *Industrial Motor Control*.

Students learn to produce wiring diagrams from schematic diagrams, connect control circuits, practice troubleshooting, analyze circuit operations, and design basic control circuits.

Each laboratory experiment requires the student to connect one or more circuits in the laboratory. The equipment used is standard motor control components such as motor starters, control relay, push buttons, and so on. The timers used in this manual are easily obtainable and can be substituted if desired. A list of the parts needed to construct each of the experiments is provided.

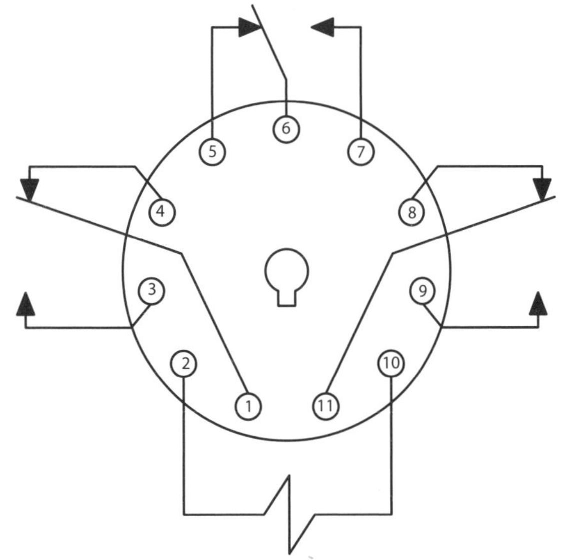

Connection diagram for an 11 pin relay

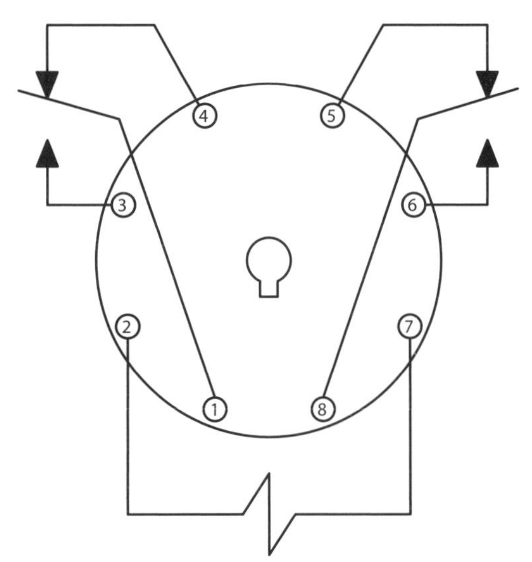

Connection diagram for an 8 pin relay

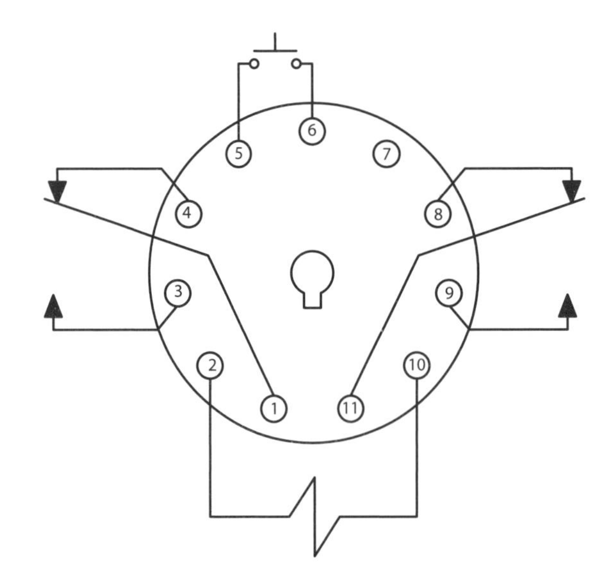

Connection diagram for a Dayton timer model 6A855
(Source: Delmar/Cengage Learning)

Typical Motor Control Lab Components

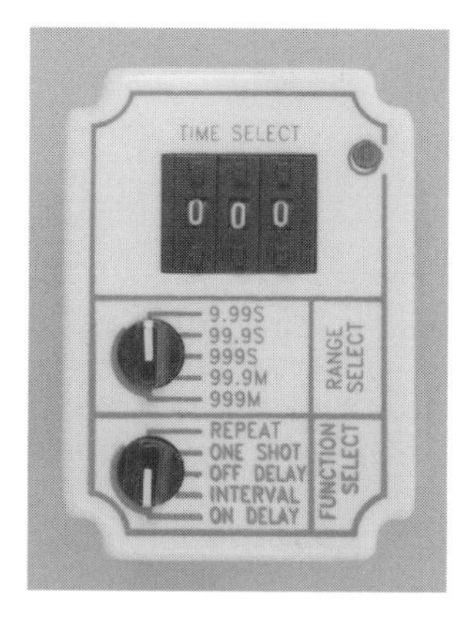

Dayton timer model 6A855. This timer mounts in an 11-pin tube socket and can be sent to operate as a repeat timer, a one-shot timer, an off-delay timer, an interval timer, and an on-delay timer. The thumb-wheel switch sets the time value. Full range times ranging from 9.99 seconds to 999 minutes can be set. (Source: Delmar/Cengage Learning)

Three-phase contactor. This contactor contains one normally open auxiliary contact and three load contacts. The contactor differs from the motor starter in that the contactor does not contain an overload relay. (Source: Delmar/Cengage Learning)

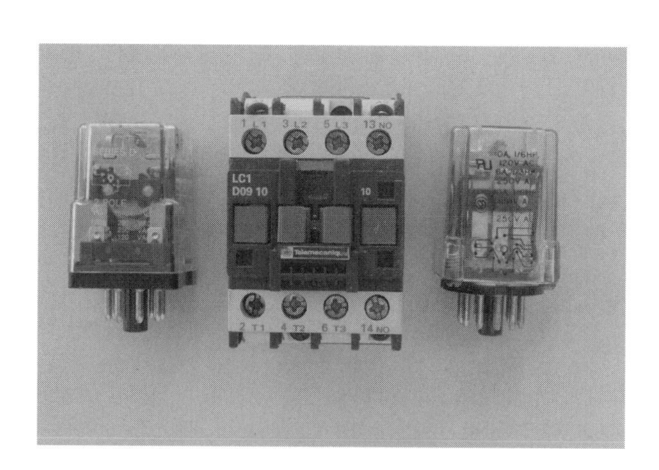

Control relays. These relays contain auxiliary contacts only and are intended to be used as part of the control circuit. They are capable of controlling low current loads such as solenoid valves, pilot lights, and the like. (Source: Delmar/Cengage Learning)

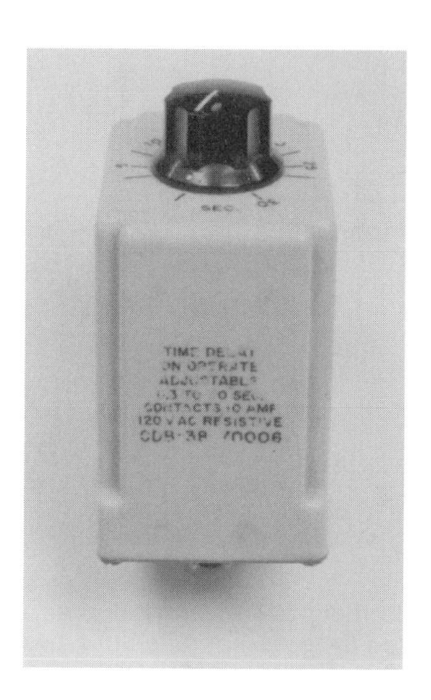

8-pin on-delay timing relay. This timer can be used as an on-delay timer only. Time setting is adjusted by the knob on top of the timer. (Source: Delmar/Cengage Learning)

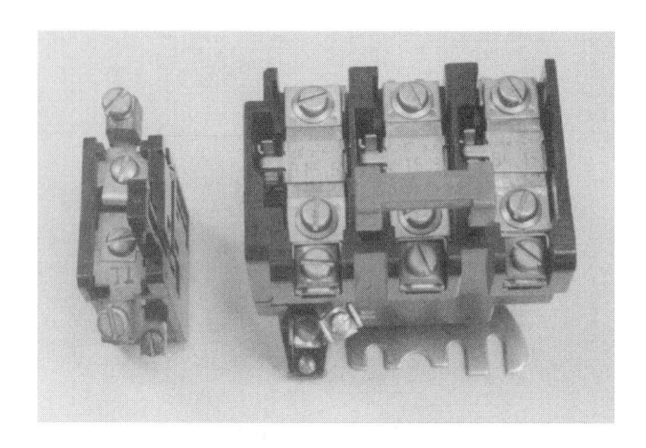

One single-phase and one three-phase overload relay. The three-phase overload relay contains three heaters, but only one set of normally closed auxiliary contacts. If an overload should occur on any of the three lines, the contacts will open. (Source: Delmar/Cengage Learning)

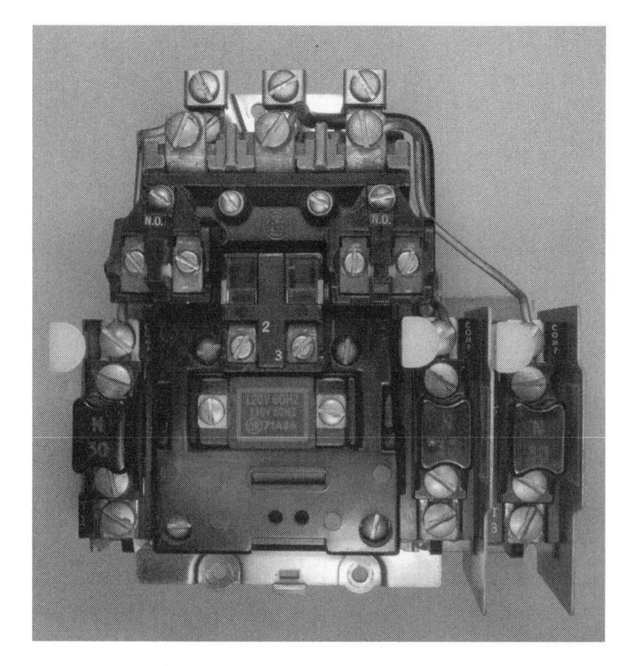

Three-phase NEMA Starter with individual OLs. This is a three-phase motor starter with three single-phase motor overload relays. The normally closed contacts of each overload relay are connected in series so that if any normally closed contact should open it will break the circuit to the motor starter coil and disconnect the motor from the line. Two sets of normally open auxiliary contacts have been added to the starter. (Source: Delmar/Cengage Learning)

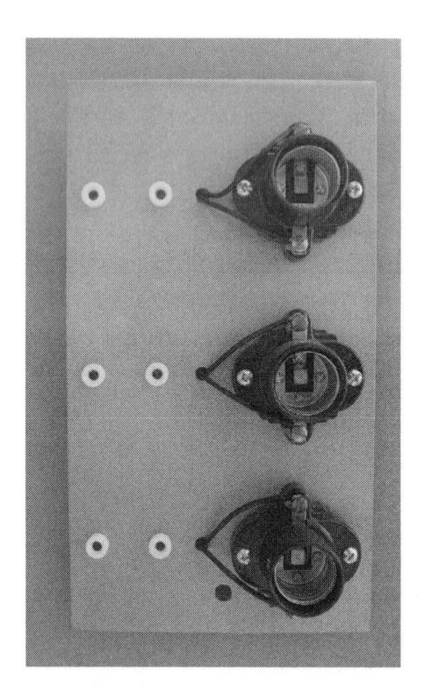

Light sockets mounted on a metal plate. Light bulbs are often used to represent different types of loads in a motor control laboratory. Mounting the sockets on a metal plate that can be hung on a rack provides easy used of the sockets. Banana jacks have been added for connection of each socket. If one end of each socket is connected together to form a wye connection, it can be used to represent a three-phase motor if one is not available. (Source: Delmar/Cengage Learning)

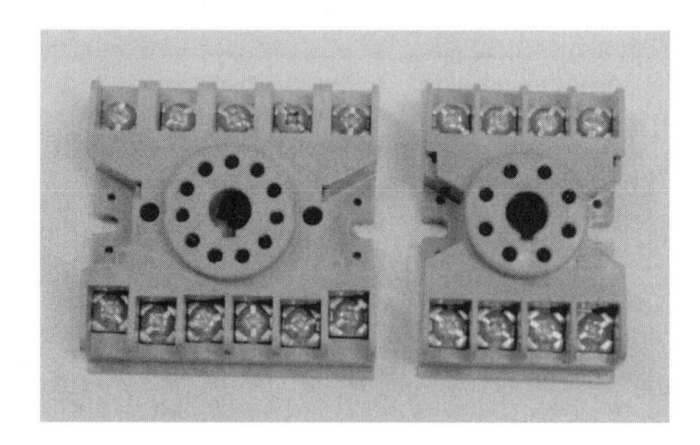

8- and 11-pin tube sockets. All wiring is done to the socket and the relay is then plugged into the socket. (Source: Delmar/Cengage Learning)

Pneumatic off-delay timer. A microswitch has been added to the bottom to supply instantaneous contacts for the timer. (Source: Delmar/Cengage Learning)

Pushbuttons mounted to a metal plate. These pushbuttons have been mounted to a metal plate that can be hung on a rack. The two top buttons are double acting in that they contain both normally open and normally closed contacts. The bottom button contains normally closed contacts only. Banana jacks have been added to each contact connection to simplify connecting circuits in the laboratory. (Source: Delmar/Cengage Learning)

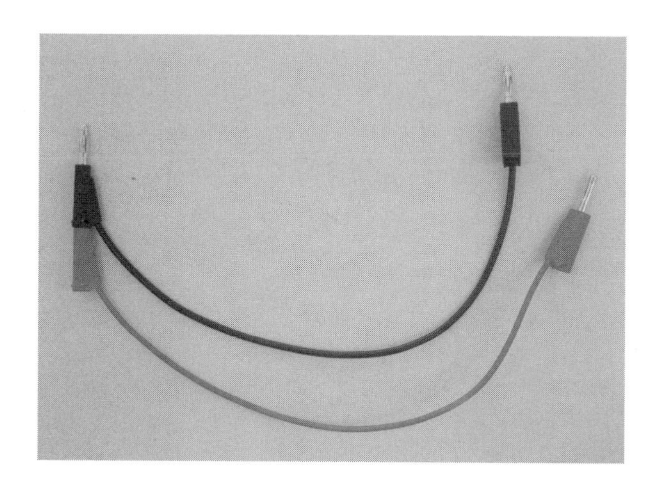

Stackable banana plugs can make connections of circuits in a laboratory quick and easy if the terminals of each control component have been equipped with banana jacks. (Source: Delmar/Cengage Learning)

Rack used for mounting motor control components. This rack is used to mount motor control components that have been placed on metal plates. The rack is an "A" frame design that permits components to be mounted on each side. Each side contains a fuse protected power supply that can provide 208/120 volts three-phase. (Source: Delmar/Cengage Learning)

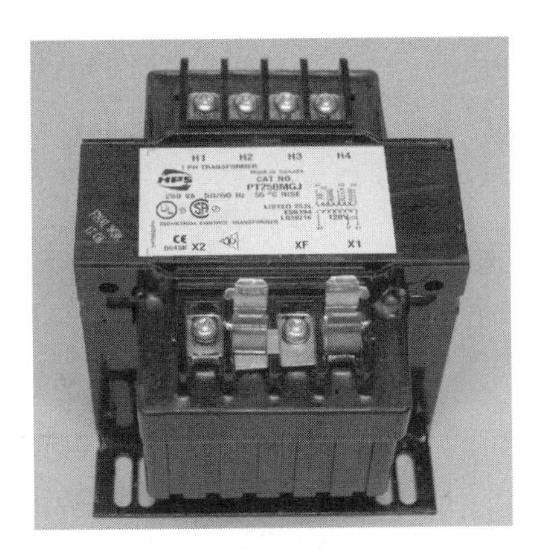

Control Transformer with tapped primary. Control transformers are used to reduce the line voltage to a value used by the control circuit. The transformer in this photo has fuse clips to permit secondary fuse protection to be added. Control transformers can be obtained that reduce different values of voltage. This transformer employs a tapped primary winding that can reduce 208, 277, or 380 volts to 120 volts. (Source: Delmar/Cengage Learning)

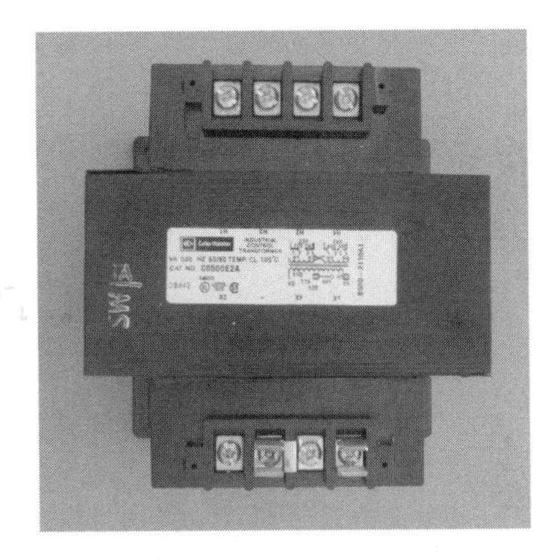

Control transformer with separate primary windings. Control transformers intended for operation on 240 or 480 volt circuits generally contain two separate primary windings. If the transformer is intended to reduce 240 volts to 120 volts, the two primary windings will be connected in parallel. If the transformer is intended to reduce 480 volts to 120 volts, the two primary windings will be connected in series. (Source: Delmar/Cengage Learning)

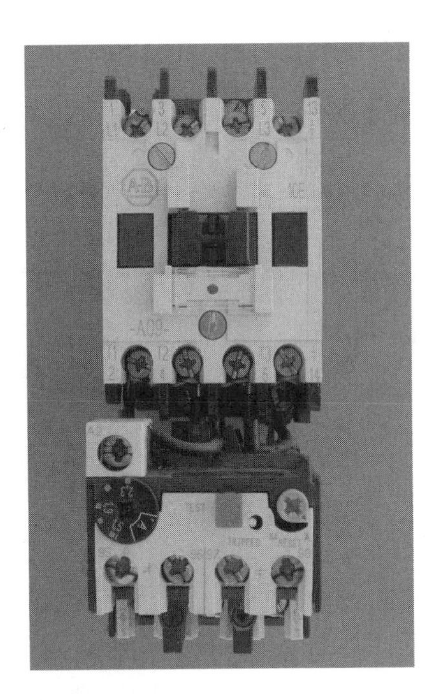

IEC type three-phase starter with three-phase overload. Note that the overload relay contains two sets of contacts instead of one. One contact is normally closed and the other is normally open. The normally closed contact is connected in series with the starter coil and the normally open contact is often used to turn on an indicator lamp to show that the motor has tripped on overload, or as the input to a programmable logic controller to inform the controller that the motor has tripped the overload relay. (Source: Delmar/Cengage Learning)

SAFETY

Name: Troy Higginbotham Date: 1-23-13 Grade:

Comments:

OBJECTIVES

After completing this worksheet, the student should be able to:

- Discuss the safety procedures for dealing with live electrical circuits.
- List common safety rules.
- Illustrate the proper procedure for testing a circuit.
- Discuss the proper procedure for energizing and de-energizing a circuit.

The experiments in this manual involve the use of line voltages of 120 volts and higher. Care must be exercised at all times when dealing with voltages of these values. Electricians who work in an industrial environment commonly work with voltages that can range from 13,800 volts to 12 volts. Some of the more common industrial voltage values are 13,800, 12,470, 7,200, 4,160, 560, 480, 277, 240, 208, 120, and 24. The typical motor control center found throughout industry (Figure S-1) can deliver enough energy under an arc-fault condition to kill a person standing 30 feet away. An arc flash will instantly set common clothing aflame. Any time a motor control circuit is to be energized or de-energized, the electrician should wear flame retardant clothing, eye protection such as a face shield or safety glasses (Figure S-2), and a hard hat (Figure S-3).

Figure S-1 Typical motor control center found in industry (Courtesy Cutler Hammer, Eaton Corp.)

Another rule to always observe when opening or closing a circuit is to never stand in front of the switch box. Always stand to one side. A short

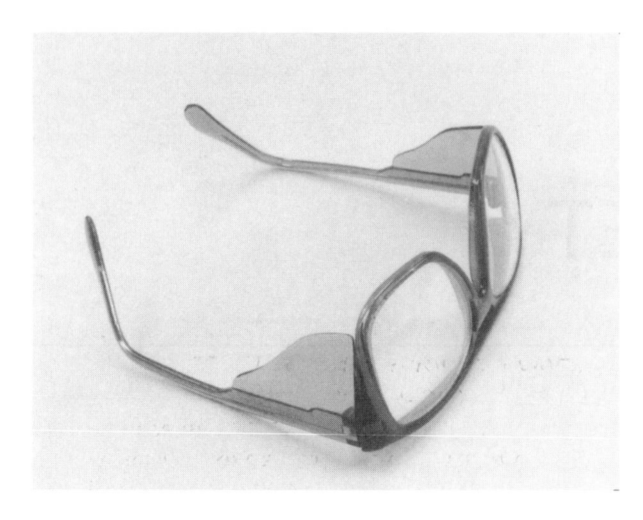

Figure S-2 Safety glasses. (Source: Delmar/Cengage Learning)

Figure S-3 Typical electrician's hard hat with attached safety goggles. (Source: Delmar/Cengage Learning)

circuit condition can cause an explosion inside the box. It is not uncommon for a shorted circuit that is not protected by the proper fuses or circuit breaker to blow the cover off the box. You do not want to be standing in front of it if that should happen.

General Safety Rules

Some general safety rules that should always be followed are:

- Never work on an energized circuit if it possible to disconnect the power.
- When checking a circuit to determine if it is energized or not, use a three-step procedure known as the test-check-test method.
  Step 1—Test the voltage measuring instrument on a known energized circuit to make certain that it is working properly.
  Step 2—Test the circuit to make certain that it is de-energized.
  Step 3—Test the voltage measuring instrument on a known energized circuit again to make certain that the instrument is still working properly.
- Think—No rule of safety is more important than this one. No amount of safeguarding or *idiot proofing* a piece of equipment can protect a person as well as taking the time to think before acting. Many technicians have been killed by supposedly "dead" circuits. Do not depend on fuses, circuit breakers, or someone else to open a circuit. If you are working on high-voltage equipment, use insulated gloves and meter probes to measure the voltage. *Think* before you touch something that could cost you your life.
- Avoid Horseplay—Jokes and horseplay have their time and place, but not when someone is working on an electric circuit or a piece of moving equipment. Do not be the cause of someone being injured or killed, and do not let someone else be the cause of your being injured or killed.
- Do Not Work Alone—This is especially true when working in a hazardous location or on a live electric circuit. Have someone with you who can turn off the power and give artificial respiration and/or cardiopulmonary resuscitation (CPR). Severe electric shock can cause breathing difficulties and can cause the heart to go into fibrillation.
- Work with One Hand When Possible—The most dangerous electric shock occurs when a current path exists from one hand to the other. This permits the current to pass directly through the heart. A person can survive a severe electric shock if the current passes between the hand and the foot but would not if the current path is through the heart.
- Learn First Aid—Anyone working on electrical equipment that operates on voltages of 50 volts or more should make an effort to learn first aid. Knowledge of first aid, especially CPR, may save your own or someone else's life.

- Avoid Alcohol and Drugs—Alcohol or drug use has no place on the work site. Alcohol and drugs are not only dangerous to the people who use them and the people who work around them, but they also cost industry millions of dollars a year. Alcohol and drug abusers kill thousands of people a year on the highway, and they are just as dangerous on a job site as they are behind the wheel of a vehicle. Many industries have instituted testing policies to screen for alcohol and drugs. A person who tests positive generally receives a warning the first time and is fired the second time.
- Use Lock-out and Tag-out Procedures—Lock-out and tag-out procedures are used to prevent someone from energizing a piece of equipment by mistake. This could apply to switches, circuit breakers, or valves. Most industries have their own internal policies and procedures. Some require that a tag similar to the one shown in Figure S-4 be placed on the piece of equipment being serviced, and some manufacturers require that the equipment be locked with a padlock. The person performing the work places the lock on the equipment and keeps the key in his or her possession. A device that permits the use of multiple padlocks and a safety tag is shown in Figure S-5. This is used when more than one person is working on the same piece of equipment. Violating lock-out and tag-out procedures is considered an extremely serious offense and in most industries is grounds for immediate termination of employment. As a general rule, there are no first-time warnings.

Figure S-4 Safety tag used to tag-out equipment. (Source: Delmar/Cengage Learning)

Figure S-5 The equipment can be locked out by several different people. (Source: Delmar/Cengage Learning)

Review Questions

1. A typical motor control center in an arc-fault condition can kill a person standing 30 feet away.

2. What is the single most important rule of safety? Think

3. List the steps for properly testing a circuit to make certain that it is de-energized.

   Test voltage on energized circuit, Test the circuit, Test voltage on energized circuit again.

4. When working with a live circuit, what is the most severe path of electric shock?

   One hand to the other

5. What is the penalty for violating lock-out and tag-out procedures in most industries?

   Termination of employment.

Section 1

STUDENT WORKSHEET

WORKSHEET 7
SERIES CIRCUITS

Name: Date: Grade:

Comments:

OBJECTIVES

After completing this worksheet, the student should be able to:

- Define a series circuit.
- Determine the total resistance of a series circuit.
- Determine the current in a series circuit.
- Determine the voltage drop across components in a series circuit.

Electric circuits can be divided into three major types: series, parallel, and combination. Series circuits are characterized by the fact that they contain only one path for current flow. In the circuit shown in Figure WS 7-1, assume that current leaves terminal L1 of the power source and returns to terminal L2. The circuit contains three components (resistors). The current must pass through each of the components because there is only one complete path from terminal L1 to terminal L2. Therefore, the current flow through components connected in series must be the same. In a series circuit, the current is the same throughout the circuit.

Because the same current must pass through each series-connected component, each offers resistance to the flow of current. The total resistance of the series circuit is therefore the sum of the resistance of each component, Figure WS 7-2. This can be expressed in the formula:

$$R_T = R_1 + R_2 + R_3 + R_N$$

In the formula, R_N means number of resistors. If a circuit contained 25 resistors connected

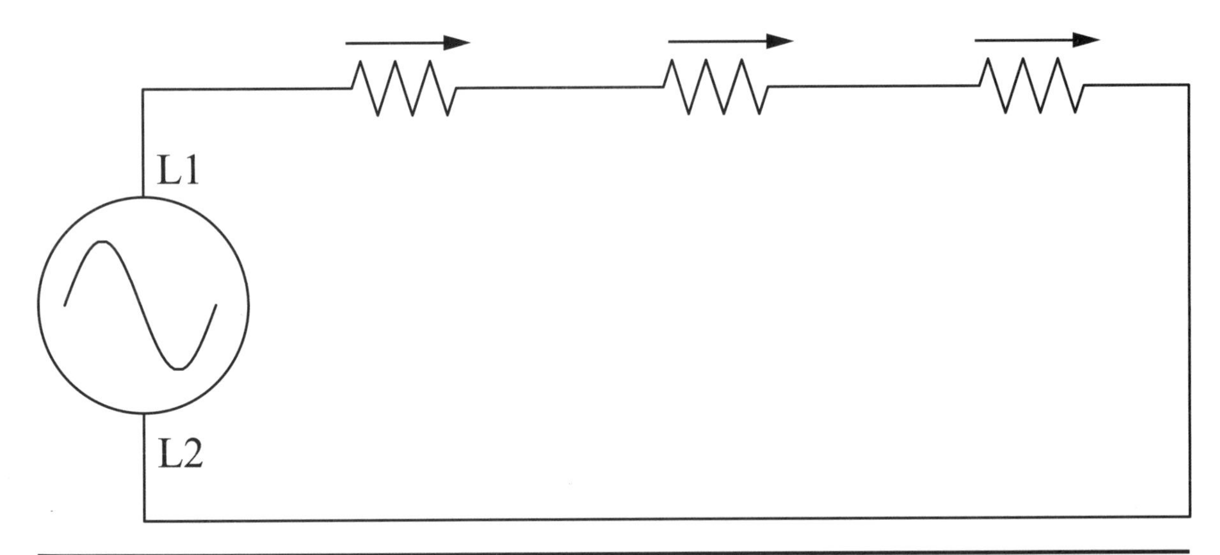

Figure WS 7-1 In a series circuit, the same current must pass through each component in the circuit. (Source: Delmar/Cengage Learning)

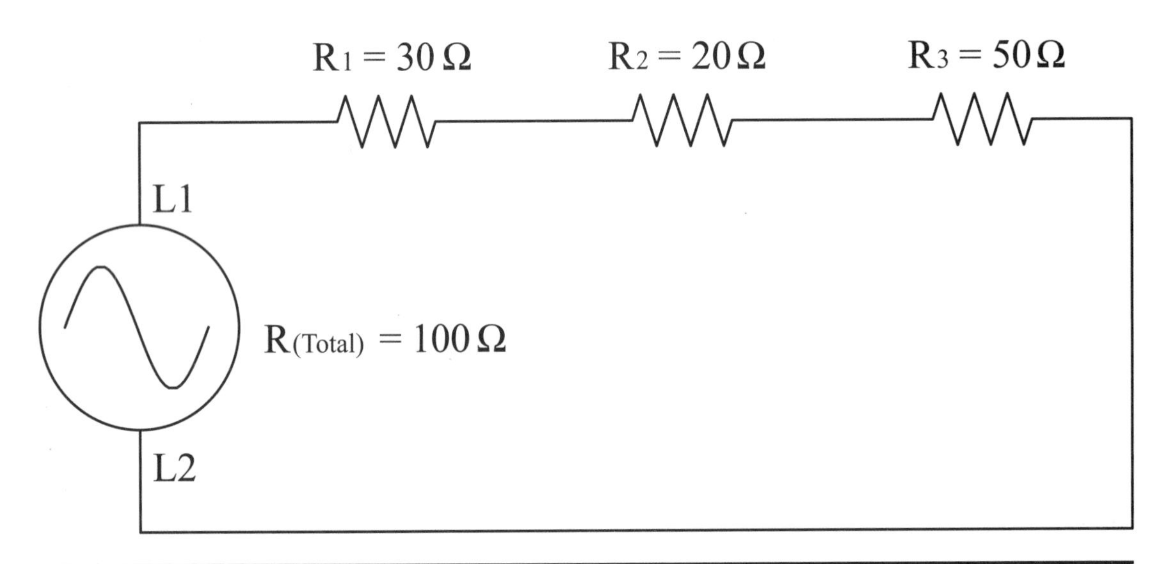

Figure WS 7-2 In a series circuit, the total resistance is the sum of the resistance of each component. (Source: Delmar/Cengage Learning)

in series, the total resistance would be the sum of R_1 through R_{25}.

The amount of voltage across each component is called its *voltage drop*. The voltage drop of a particular component is the portion of the total circuit voltage (applied voltage) necessary to cause the circuit current to flow through that amount of resistance. Assume that the circuit shown in Figure WS 7-3 has an applied or a total voltage of 120 volts. Because the total resistance and total voltage are known, the circuit current can be computed using Ohm's law.

$$I = \frac{E}{R}$$

$$I = \frac{120}{100}$$

$$I = 1.2 \text{ A}$$

Now that the circuit current is known, the voltage drop across each component can be computed using Ohm's law.

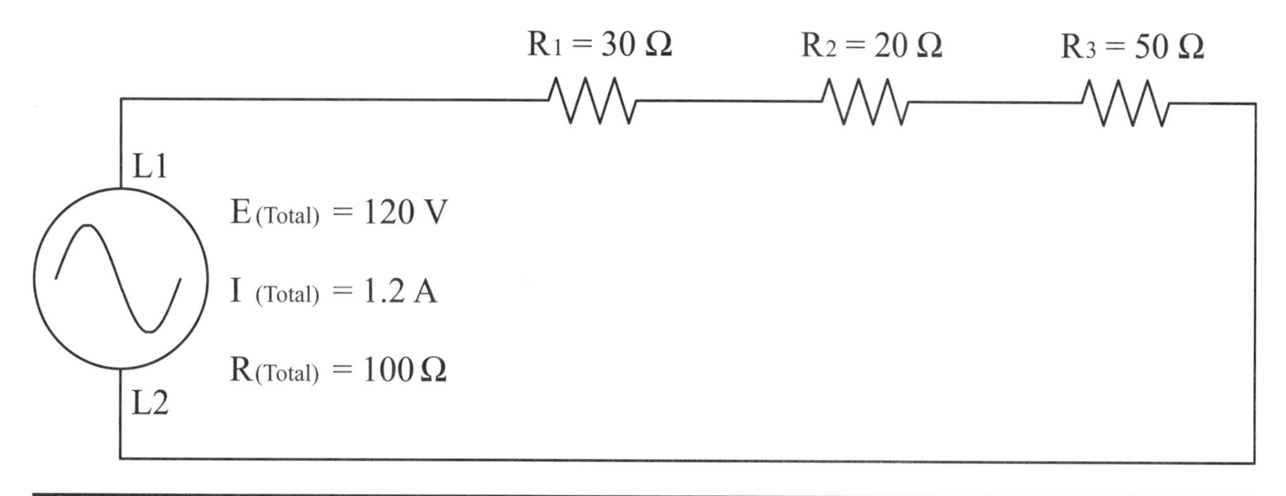

Figure WS 7-3 Since total resistance and total voltage are known, the total current can be computed using Ohm's Law. (Source: Delmar/Cengage Learning)

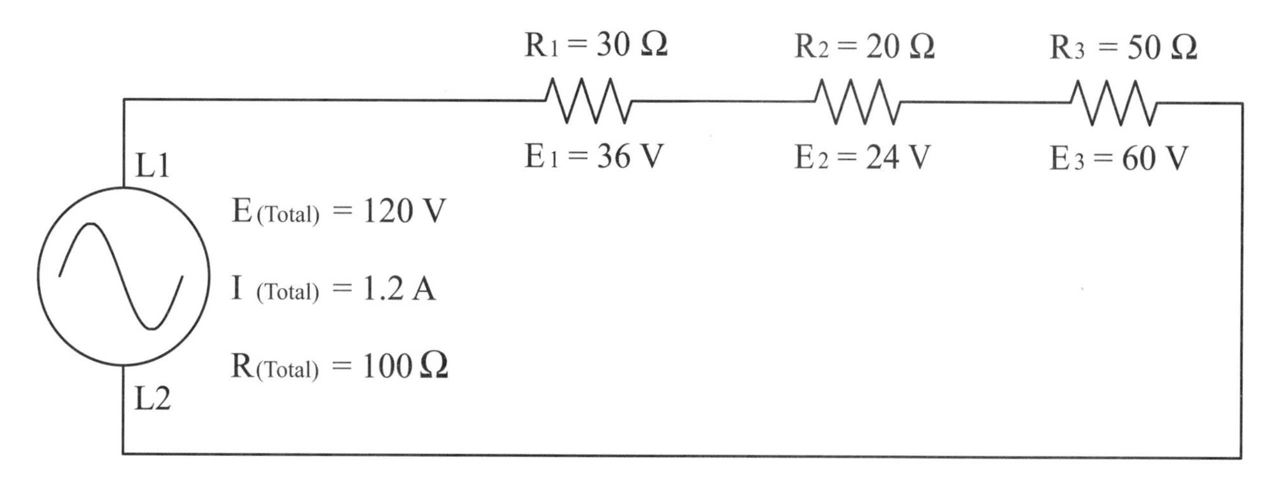

Figure WS 7-4 The sum of the voltage drops across each component must equal the applied voltage. (Source: Delmar/Cengage Learning)

$E_1 = I \times R_1$

$E_1 = 1.2 \times 30$

$E_1 = 36$ V

$E_2 = I \times R_2$

$E_2 = 1.2 \times 20$

$E_2 = 24$ V

$E_2 = I \times R_2$

$E_2 = 1.2 \times 50$

$E_2 = 60$ V

Notice that the sum of the voltage drops across each component in Figure WS 7-4 is equal to the total or applied voltage. There are three basic rules that can be applied to a series circuit:

1. The current is the same in any part of the circuit. This rule can be simply stated as: *Current remains the same.*
2. The total resistance is the sum of the individual resistive components. A simple way of stating this is: *Resistance adds.*
3. The sum of the voltage drop across each component must equal the applied circuit voltage. A simpler way of stating this rule is: *Voltage drops add.*

These three rules and Ohm's law can be used to determine circuit values in a series circuit when other values are known. In the following illustrations, find the missing value using the three rules for series circuits and Ohm's law. State the rule or rules employed in the problem and the formula or formulas used to find the answer.

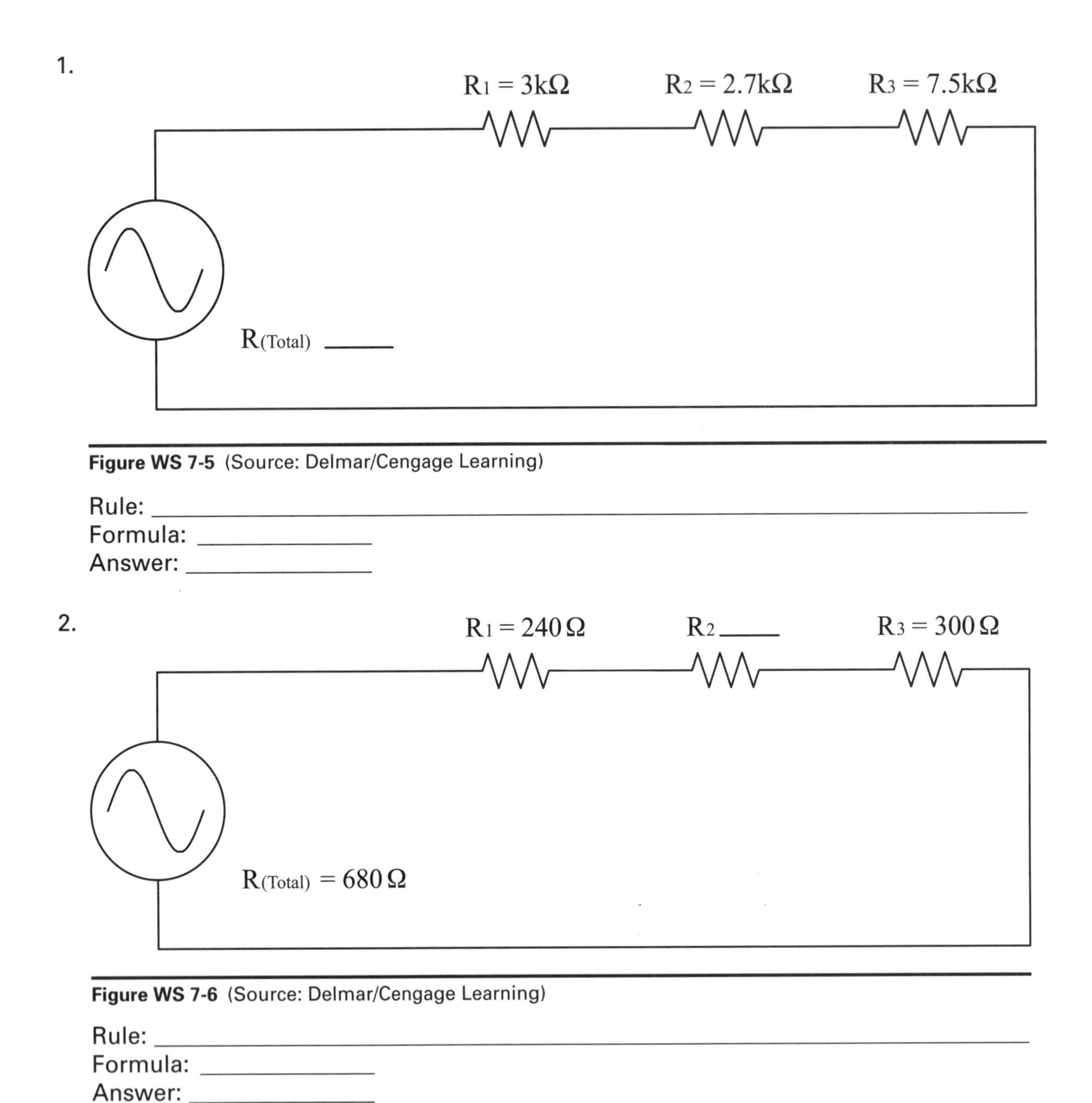

1.

Figure WS 7-5 (Source: Delmar/Cengage Learning)

Rule: ____________________

Formula: ____________

Answer: ____________

2.

Figure WS 7-6 (Source: Delmar/Cengage Learning)

Rule: ____________________

Formula: ____________

Answer: ____________

3.

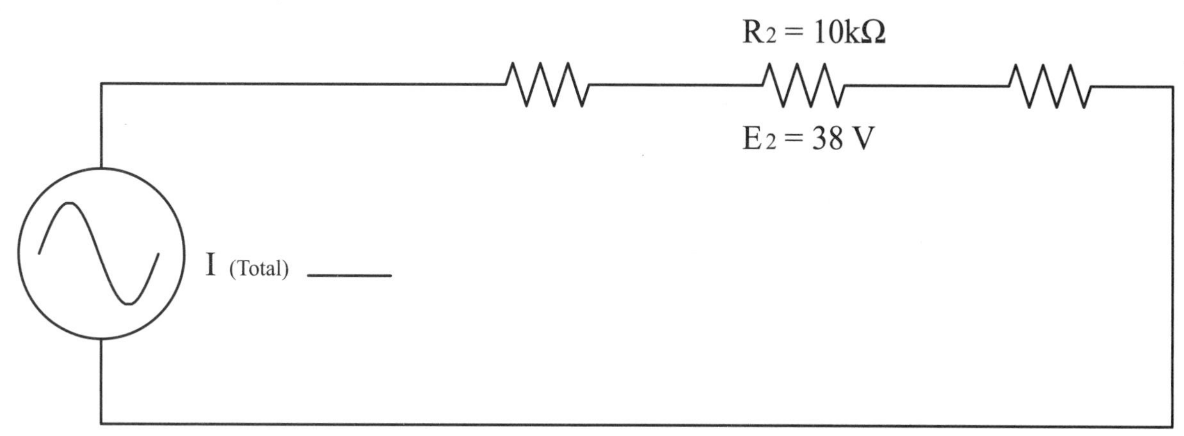

Figure WS 7-7 (Source: Delmar/Cengage Learning)

Rule: ____

Formula: ______________

Answer: ______________

4.

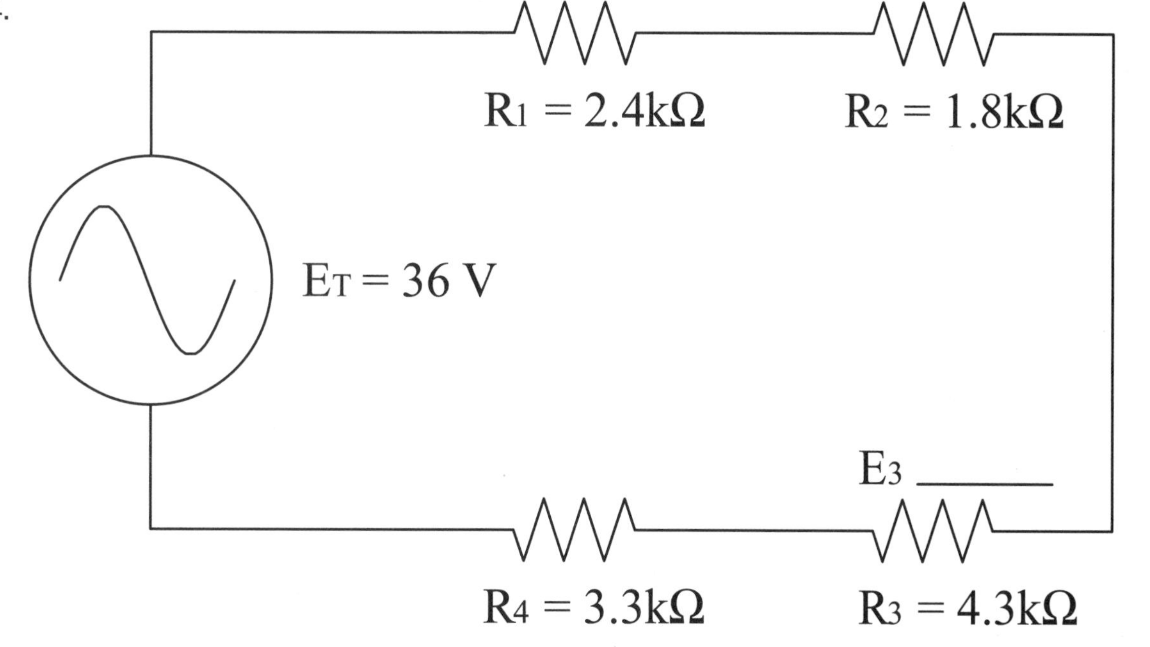

Figure WS 7-8 (Source: Delmar/Cengage Learning)

Rule: ____

Formula: ______________

Answer: ______________

5.

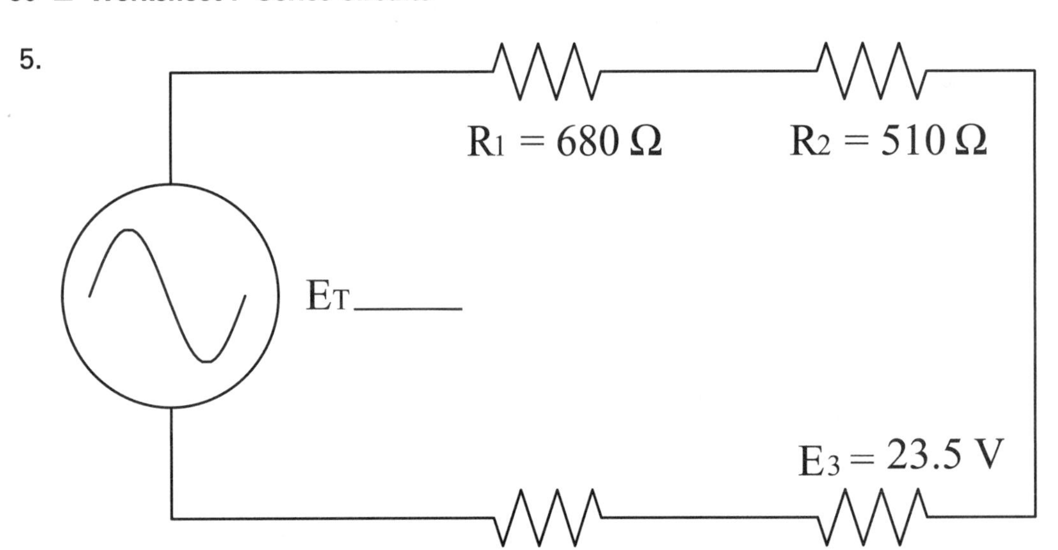

Figure WS 7-9 (Source: Delmar/Cengage Learning)

Rule: ____

Formula: ______________

Answer: ______________

6.

$R_1 = 91k\Omega$

E_1 12.6 Volts

$R_4 = 68k\Omega$

E_4 ______

Figure WS 7-10 (Source: Delmar/Cengage Learning)

Rule: ____

Formula: ______________

Answer: ______________

7.

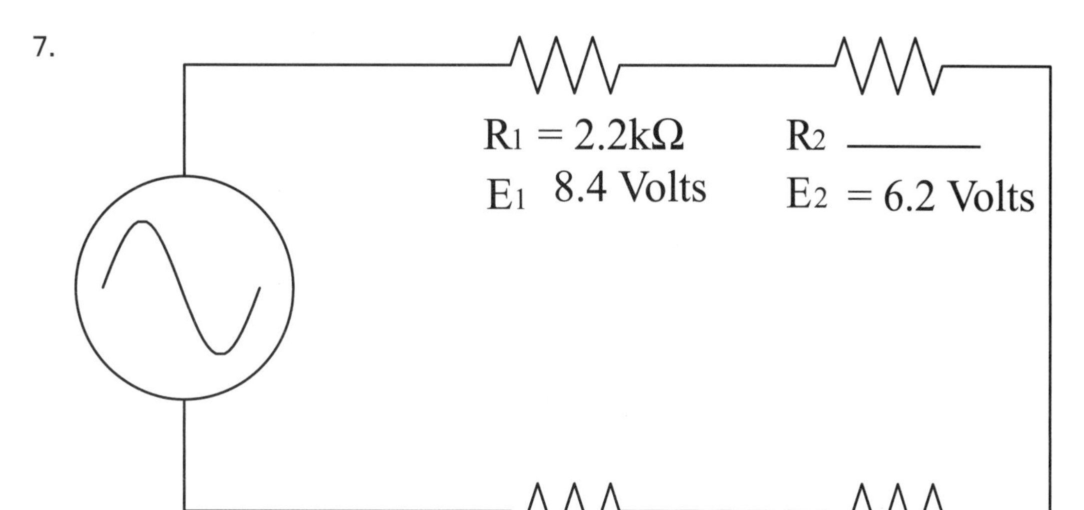

Figure WS 7-11 (Source: Delmar/Cengage Learning)

Rule: ______________________________

Formula: ______________

Answer: ______________

8.

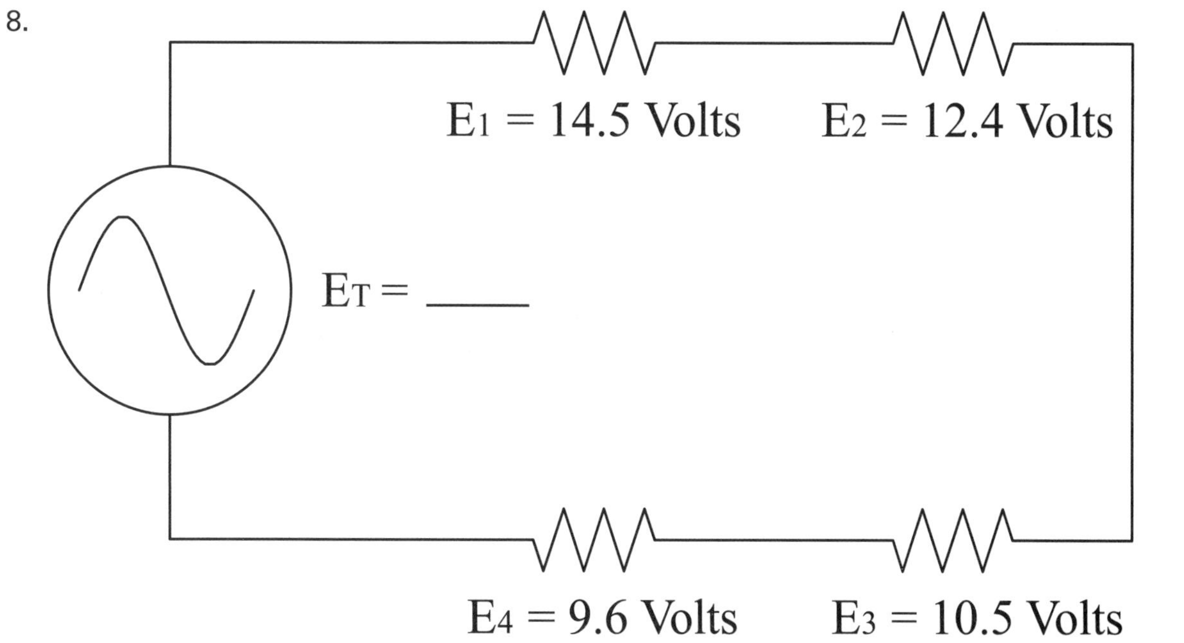

Figure WS 7-12 (Source: Delmar/Cengage Learning)

Rule: ______________________________

Formula: ______________

Answer: ______________

9.

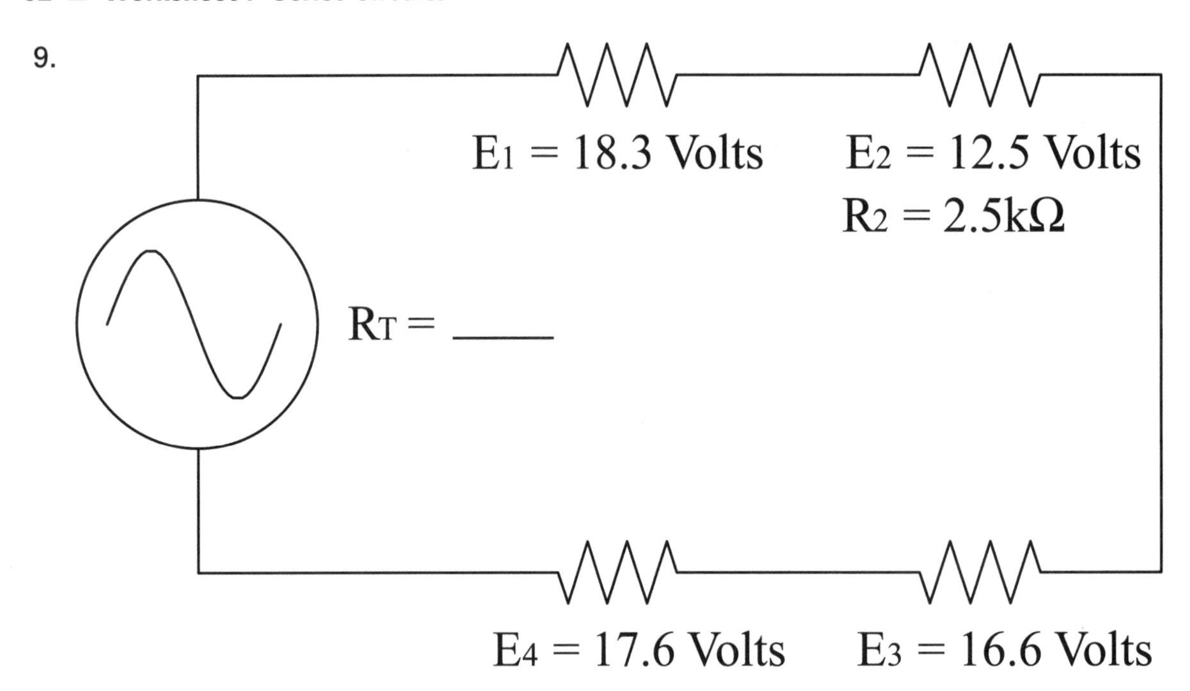

Figure WS 7-13 (Source: Delmar/Cengage Learning)

Rule: ____

Formula: ____________

Answer: ____________

10.

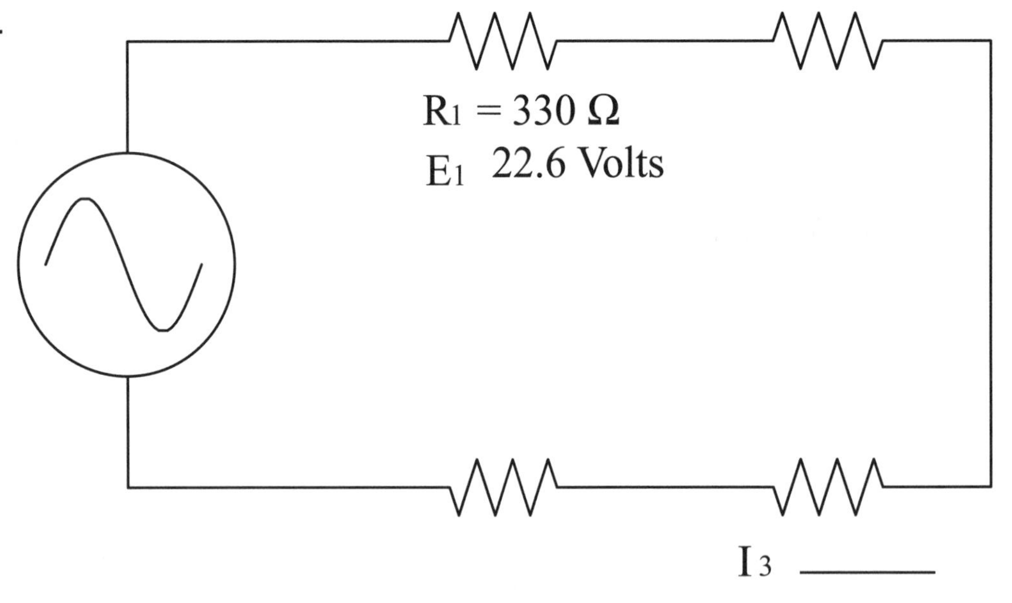

Figure WS 7-14 (Source: Delmar/Cengage Learning)

Rule: ____

Formula: ____________

Answer: ____________

WORKSHEET 8
PARALLEL CIRCUITS

Name:	Date:	Grade:
Comments:		

OBJECTIVES

After completing this worksheet, the student should be able to:

- Define a parallel circuit.
- Determine the total resistance of a parallel circuit.
- Determine the current in a parallel circuit.
- Calculate values of voltage, current, and resistance in a parallel circuit.

Parallel circuits contain more than one path for current flow, Figure WS 8-1. Assume that current leaves the power source at point A and must return to point B. In the circuit shown, there are three different paths for current flow. Current can leave the power source and return through resistor R_1. It can also return through resistor R_2 or resistor R_3. One of the rules concerning parallel circuits is that *the total current in the circuit is the sum of the currents that flow through each branch.* Ammeters have been added to the circuit shown in Figure WS 8-2. Six amperes of current leave the power source at point A: 1 ampere of current flows through resistor R_1, and 5 amperes continue to resistors R_2 and R_3; 2 amperes of current flow through resistor R_2, leaving 3 amperes to flow through R_3. The current returns to point B of the power source in the same manner.

A second rule of parallel circuits is that the *voltage is the same across all branches*, Figure WS 8-3. Assume that the power source is supplying a voltage of 120 volts. If a voltmeter is connected across each branch, it will show that the same voltage is applied

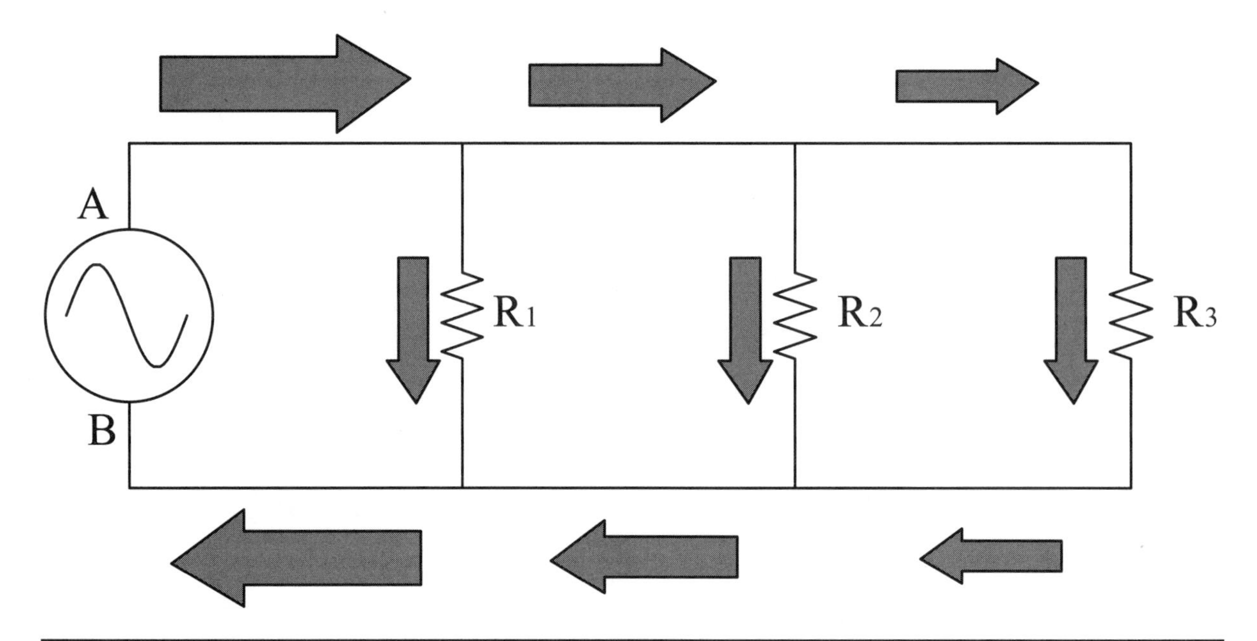

Figure WS 8-1 Parallel circuits contain more than one path for current flow. (Source: Delmar/Cengage Learning)

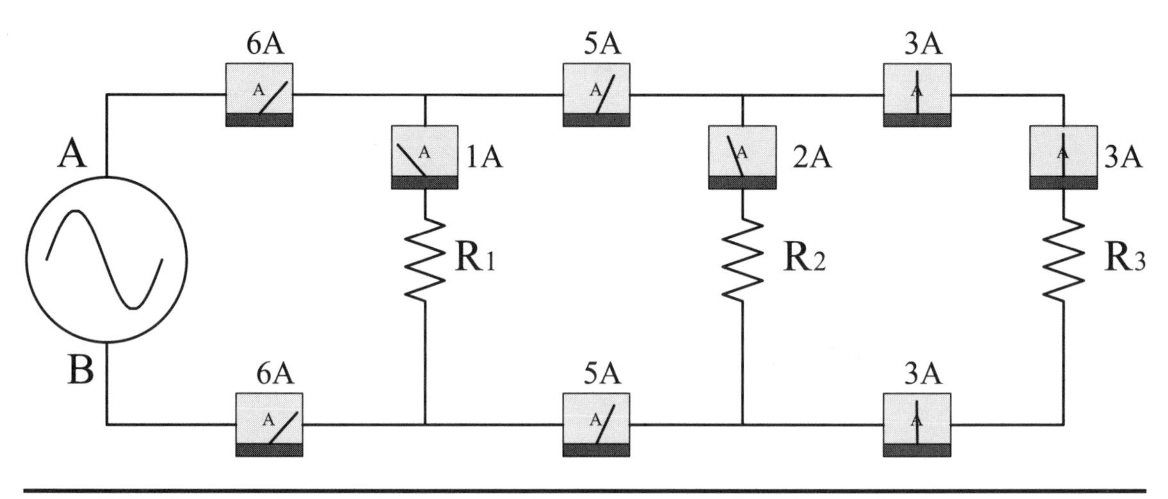

Figure WS 8-2 The total current is the sum of the currents through each branch. (Source: Delmar/Cengage Learning)

across each branch. Lighting and receptacle circuits in homes are connected in parallel. In this way, each lamp or receptacle receives the same voltage.

In a parallel circuit, the total resistance is always less than the resistance of any single branch. This is due to the fact that each time another branch is added to the circuit another path for current flow is created, thereby reducing the total resistance to the flow of current. This often causes confusion for some because it leads to the idea that resistance is load. You can add load to a circuit by connecting another device or resistor to the circuit. If that device is connected in parallel, however, the total resistance is decreased, not increased.

Three formulas can be used for determining the total resistance of a circuit. The first formula is employed when all resistance values in a parallel circuit are the same, Figure WS 8-4. The formula states that the total resistance can

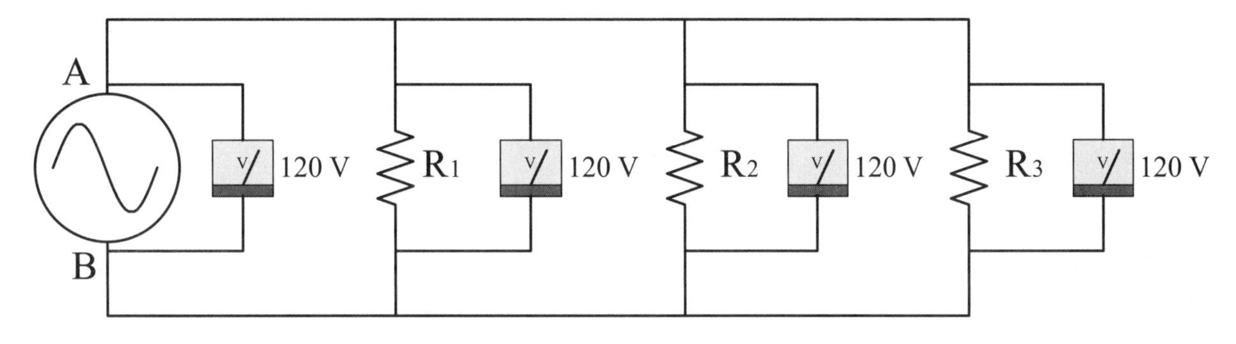

Figure WS 8-3 The voltage is the same across all branches of a parallel circuit. (Source: Delmar/Cengage Learning)

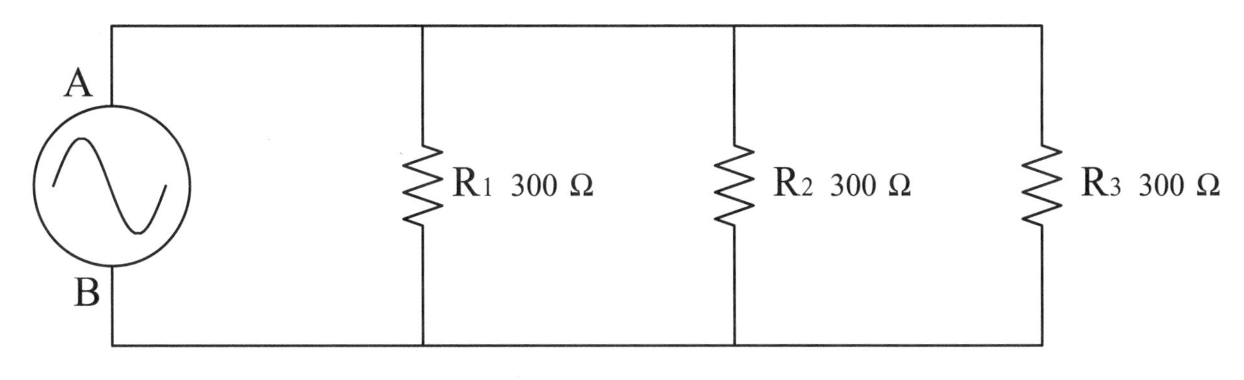

Figure WS 8-4 All resistors have the same value. (Source: Delmar/Cengage Learning)

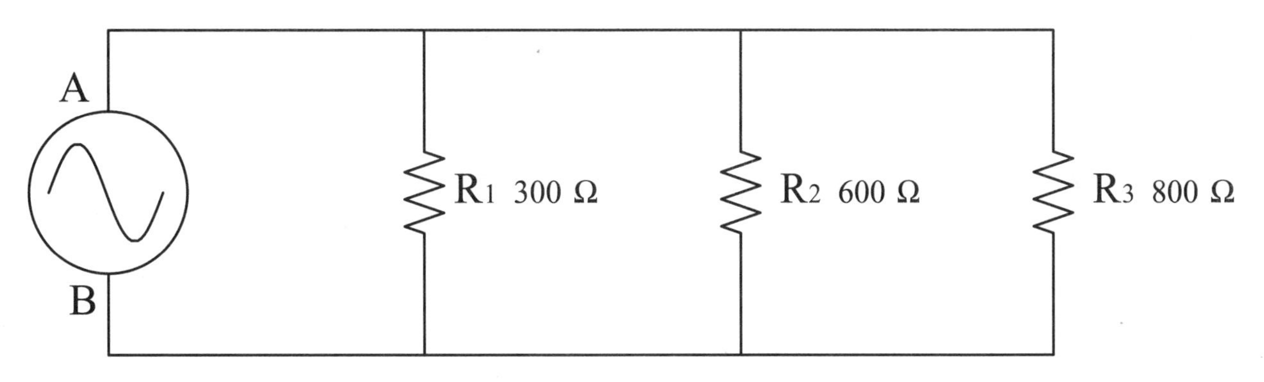

Figure WS 8-5 Three resistors with different values are connected in parallel. (Source: Delmar/Cengage Learning)

be determined by dividing the value of one resistor by the number of resistors.

$$R_T = \frac{R}{N}$$

$$R_T = \frac{300}{3}$$

$$R_T = 100\ \Omega$$

The second formula is called the product over sum formula. This formula can be employed to find the combined resistance value of two resistors at a time. In the circuit shown in Figure WS 8-5, a 300 Ω resistor, a 600 Ω resistor, and an 800 Ω resistor are connected in parallel. It is assumed that the 300 Ω resistor is R_1, the 600 Ω resistor is R_2, and the 800 Ω resistor is R_3.

$$R_T = \frac{R_1 \times R_2}{R_1 \times R_2}$$

$$R_T = \frac{300 \times 600}{300 + 600}$$

$$R_T = \frac{180{,}000}{900}$$

$$R_T = 200\ \Omega$$

The first two resistors have a combined resistance of 200 Ω. The 300 Ω and 600 Ω resistors can be replaced with a single resistor with a value of 200 Ω, Figure WS 8-6. The combined resistor is connected in parallel with an 800 Ω resistor. The combined resistor will now be considered as R_1 in the formula and the 800 Ω resistor will be considered as R_2.

$$R_T = \frac{200 \times 800}{200 + 800}$$

$$R_T = \frac{160{,}00}{1000}$$

$$R_T = 160\ \Omega$$

The total resistance of the circuit is 160 Ω. If a fourth resistor were connected in parallel, the process would be repeated using 160 Ω as R_1 in the formula and the value of the fourth resistor as R_2.

The third formula is probably the most used because of scientific calculators. This formula is called the reciprocal formula. A reciprocal is any number divided into 1. The reciprocal formula states that *the reciprocal of the total resistance is equal to the sum of the reciprocals of each branch.*

$$\frac{1}{R_T} = \frac{1}{R_1} + \frac{1}{R_2} + \frac{1}{R_2} + \frac{1}{R_N}$$

This formula can be modified so that it will find the total resistance, not the reciprocal of the total resistance.

$$R_T = \frac{1}{\frac{1}{R_1} + \frac{1}{R_2} + \frac{1}{R_3} + \frac{1}{R_N}}$$

To use this formula, place the numeric resistive values into the formula. In this example, the circuit contains three resistors.

$$R_T = \frac{1}{\frac{1}{300} + \frac{1}{600} + \frac{1}{800}}$$

Scientific-type calculators contain a reciprocal key (1/x). Any number on the display of a calculator is on the X axis. The reciprocal key will automatically divide any number on the display into 1. To calculate the total resistance, input the following on your calculator:

[3] [0] [0] [1/x] [+] [6] [0] [0] [1/x] [+]

[8] [0] [0] [1/x] [=] [1/x]

The formula can be modified to determine the value of a parallel-connected resistor if the total resistance is known, Figure WS 8-7.

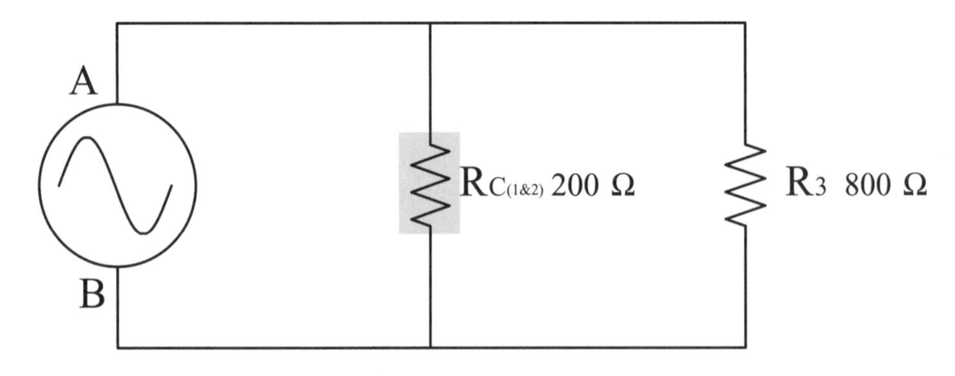

Figure WS 8-6 The first two resistors have been combined to form one resistor. (Source: Delmar/Cengage Learning)

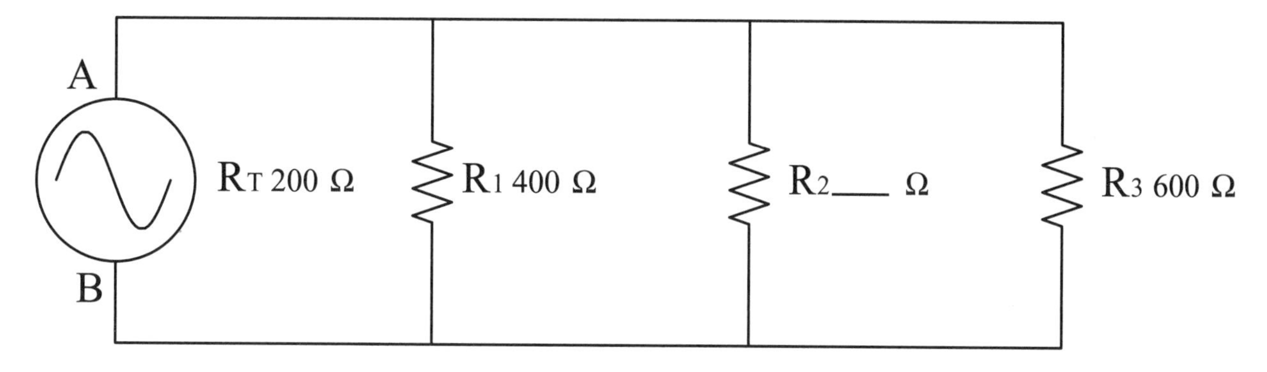

Figure WS 8-7 The second resistor value is unknown. (Source: Delmar/Cengage Learning)

$$R_2 = \frac{1}{\frac{1}{R_T} - \left(\frac{1}{R_1} + \frac{1}{R_2}\right)}$$

$$R_2 = \frac{1}{\frac{1}{200} - \left(\frac{1}{400} + \frac{1}{600}\right)}$$

To determine the value of resistor R_2, enter the following on your calculator:

[2] [0] [0] [1/x] [–] [(] [4] [0] [0] [+]

[6] [0] [0] [1/x] [)] [=] [1/x]

$$R_2 = 1{,}200\ \Omega$$

In the following examples, determine the missing value. List the formula or formulas used to determine the answer.

1.

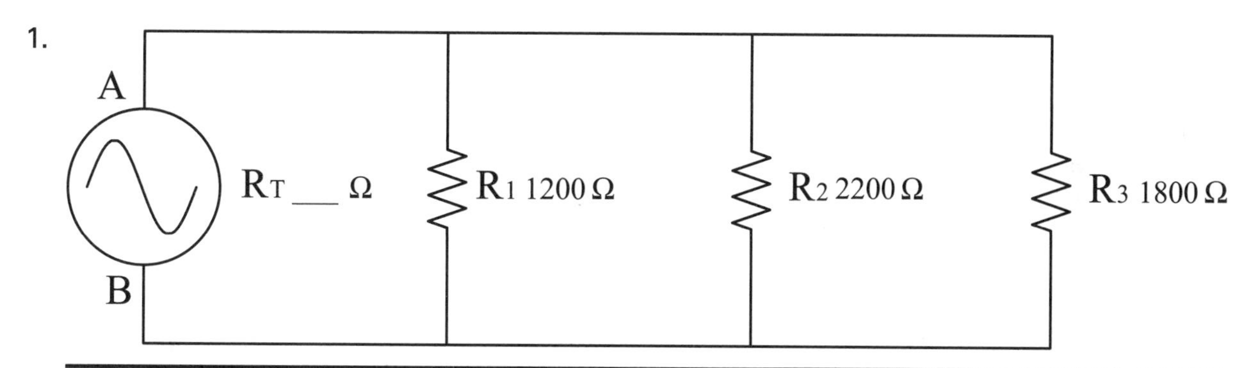

Figure WS 8-8 (Source: Delmar/Cengage Learning)

Formula: ______________

Answer: ______________

2.

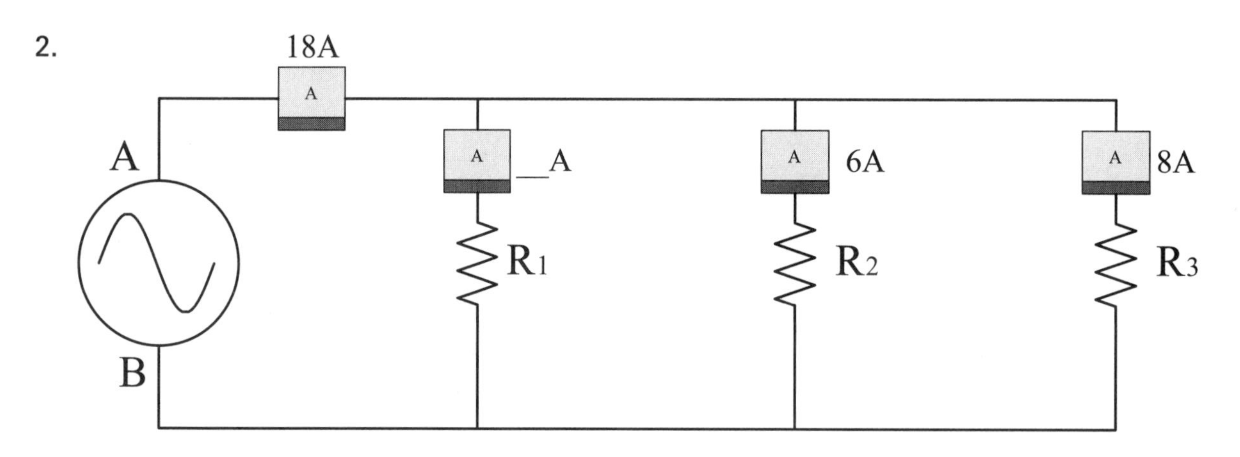

Figure WS 8-9 (Source: Delmar/Cengage Learning)

Formula: ______________

Answer: ______________

3.

A
1.5A
0.75A
R1
R2 20 Ω
R3 __
B

Figure WS 8-10 (Source: Delmar/Cengage Learning)

Formula: ______________

Answer: ______________

4.

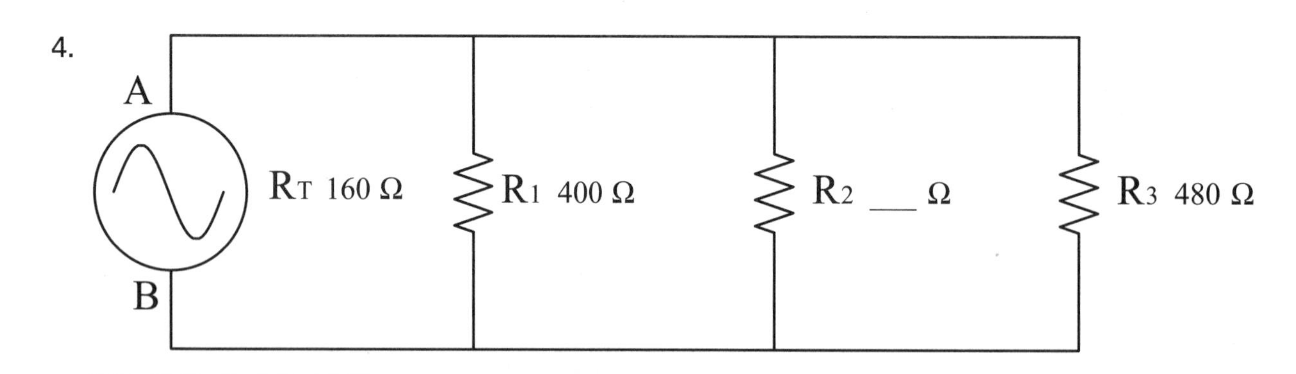

Figure WS 8-11 (Source: Delmar/Cengage Learning)

Formula: ______________

Answer: ______________

5.

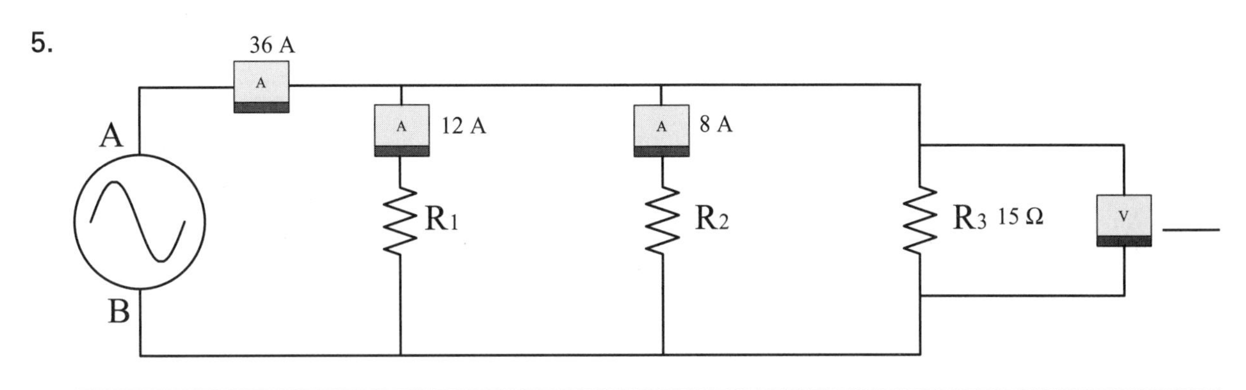

Figure WS 8-12 (Source: Delmar/Cengage Learning)

Formula: ______________

Answer: ______________

6.

A

RT ___ Ω

R1 5.1kΩ

R2 4.7kΩ

R3 7.5kΩ

B

Figure WS 8-13 (Source: Delmar/Cengage Learning)

Formula: ______________

Answer: ______________

7.

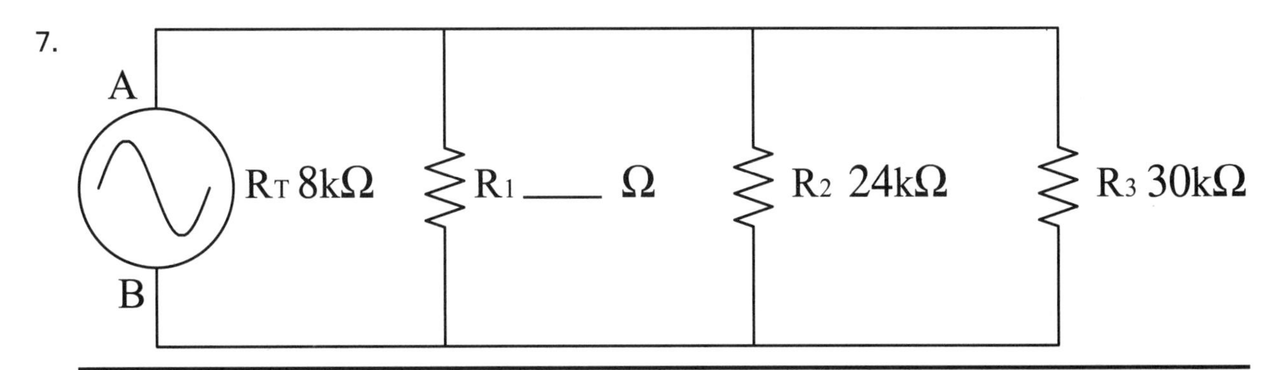

Figure WS 8-14 (Source: Delmar/Cengage Learning)

Formula: ______________

Answer: ______________

8.

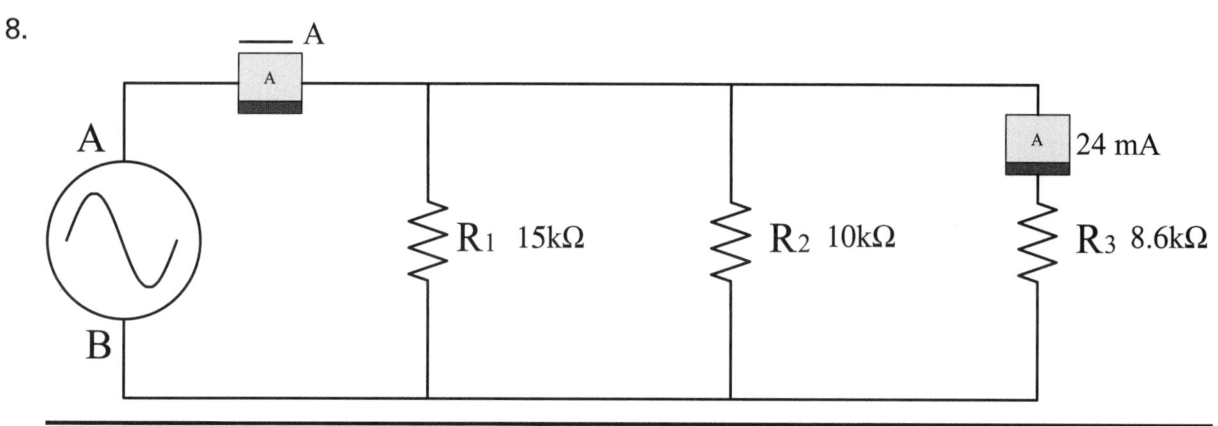

Figure WS 8-15 (Source: Delmar/Cengage Learning)

Formula: ______________
Answer: ______________

9.

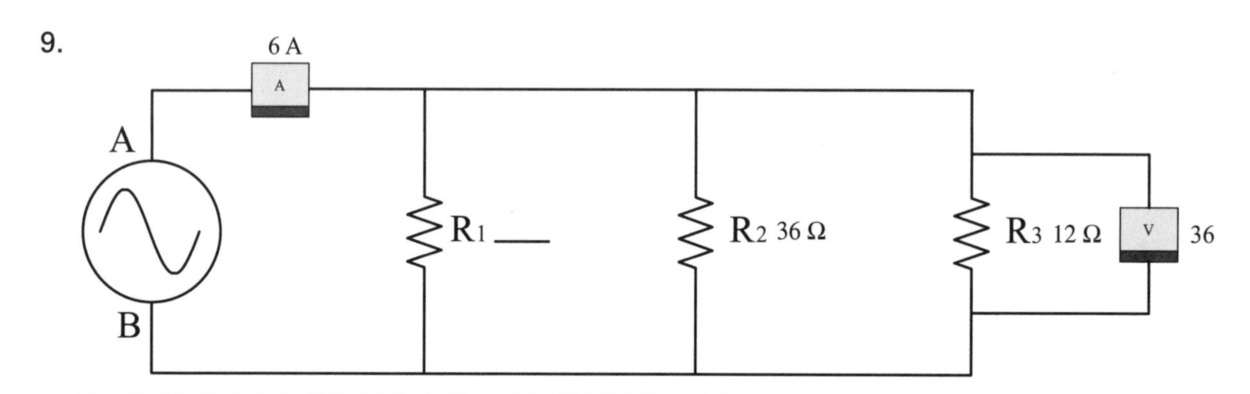

Figure WS 8-16 (Source: Delmar/Cengage Learning)

Formula: ______________
Answer: ______________

10.

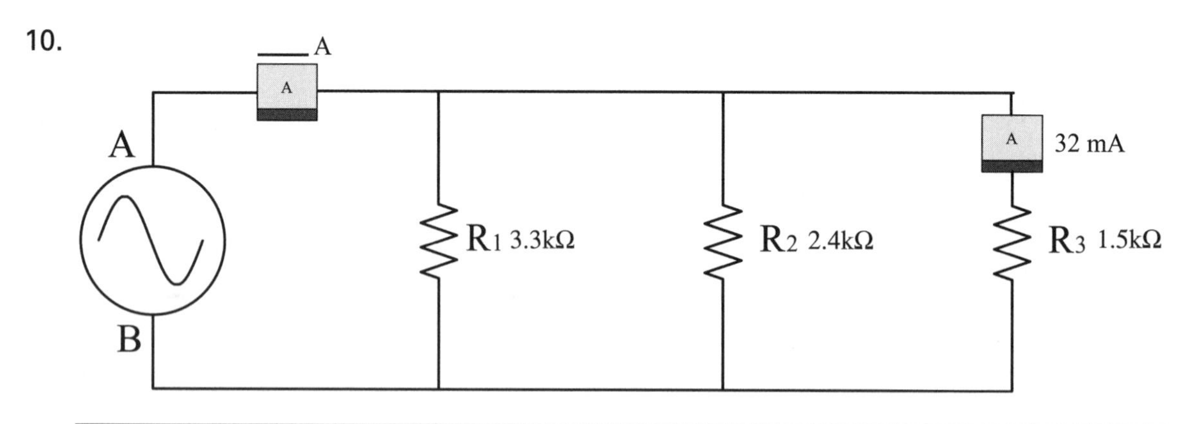

Figure WS 8-17 (Source: Delmar/Cengage Learning)

Formula: ______________
Answer: ______________

WORKSHEET 9

COMBINATION CIRCUITS

Name: Date: Grade:

Comments:

OBJECTIVES

After completing this worksheet, the student should be able to:

- Reduce a combination circuit to a simple series or parallel circuit.
- Determine values of voltage, current, and resistance in a combination circuit.

Combination circuits contain both series and parallel parts, Figure WS 9-1. Resistors R_2 and R_3 are connected in parallel with each other. Resistor R_1 is connected in series with resistors R_2 and R_3. The simplest way to determine which parts of a circuit are connected in parallel with each other and which parts are connected in series is to trace the current path through the circuit. Assume that current leaves the top of the alternator and must return to the bottom. All of the circuit current must flow through resistor R_1. The definition of a series circuit states that there is only one path for current flow. Resistor R_1, therefore, is connected in series with the rest of the circuit. After the current passes through resistor R_1, it can divide between resistors R_2 and R_3. The definition of a parallel circuit states that current can flow through more than one path. Resistors R_2 and R_3, therefore, are connected in parallel with each other. Values have been added to resistors R_1, R_2, and R_3 in Figure WS 9-2, The best way to determine the total resistance of the circuit is to combine parts of the circuit to form a simpler

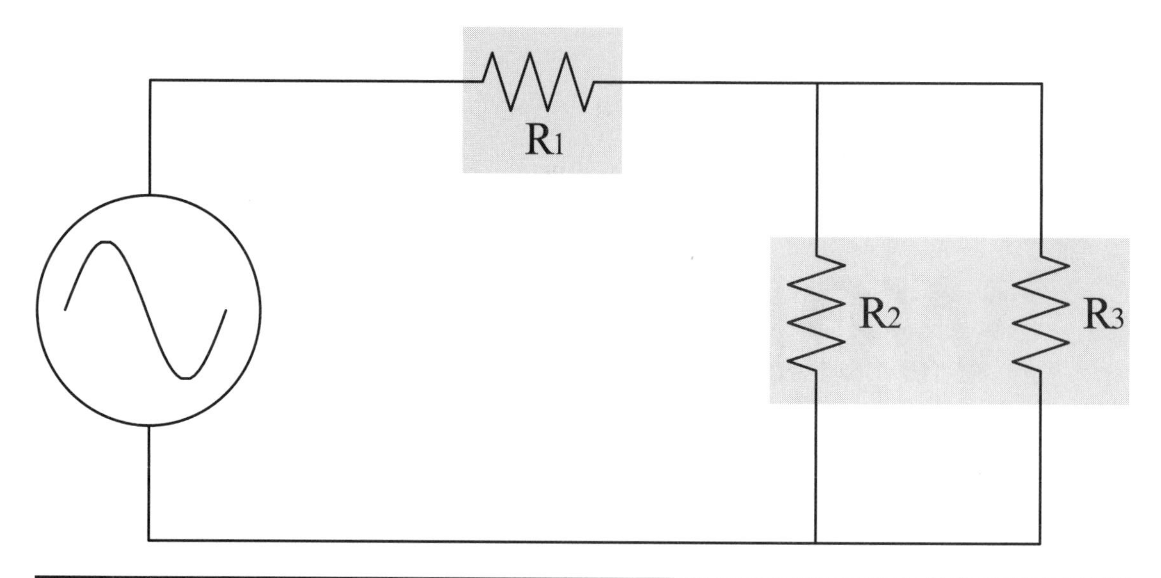

Figure WS 9-1 Combination circuits contain both series and parallel parts. (Source: Delmar/ Cengage Learning)

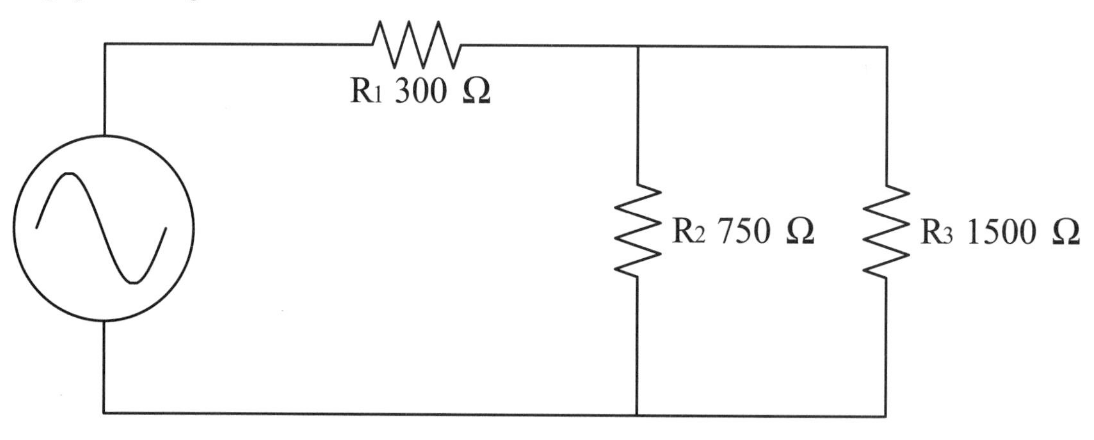

Figure WS 9-2 Resistor values are added to the circuit. (Source: Delmar/Cengage Learning)

circuit. Because resistors R_2 and R_3 are connected in parallel, they can be reduced to a single resistor by finding their total value.

$$R_T = \frac{1}{\frac{1}{R_1} + \frac{1}{R_2} + \frac{1}{R_3} + \frac{1}{R_N}}$$

$$R_T = \frac{1}{\frac{1}{750} + \frac{1}{1500}}$$

$$R_T = 500\ \Omega$$

Resistors R_2 and R_3 can now be replaced with a single resistor designated as $R_{C(2\text{-}3)}$ (resistor combination 2 and 3) with a value of 500 Ω as shown in Figure WS 9-3. The circuit has now become a simple series circuit with two resistors having values of 300 Ω and 500 Ω. It is also assumed that the circuit has an applied voltage of 120 volts. Because the circuit is now a series circuit, the total resistance can be determined by adding the resistor values together.

$$R_T = R_T + R_{c(2-3)}$$

$$R_T = 300 + 500$$

$$R_T = 800\ \Omega$$

Now that the total resistance is known, the circuit current can be determined using Ohm's law.

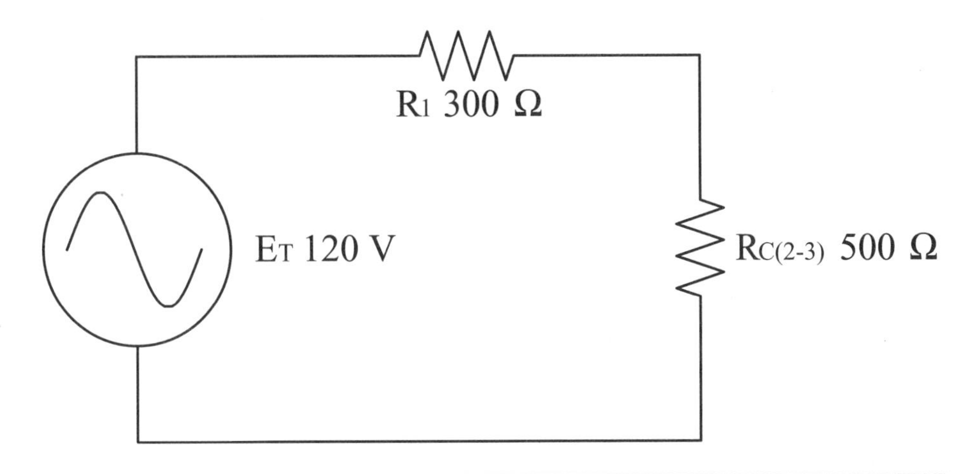

Figure WS 9-3 Resistors R_2 and R_3 have been combined to make one resistor. (Source: Delmar/Cengage Learning)

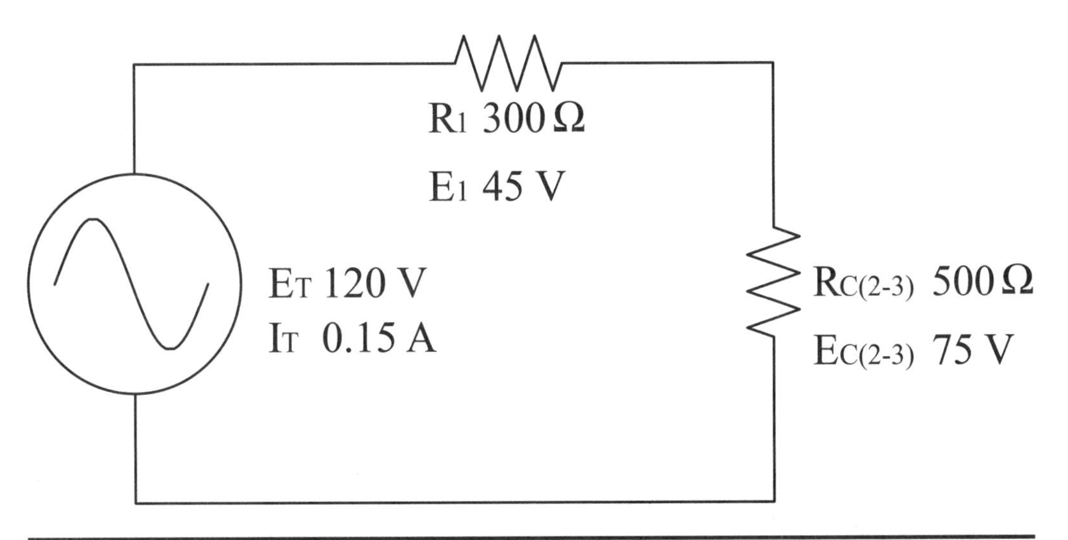

Figure WS 9-4 The voltage drop across each component can now be determined. (Source: Delmar/Cengage Learning)

$$I_T = \frac{E_T}{R_T}$$

$$I_T = \frac{120}{800}$$

I_T = 0.15 ampere or 150 mA

Because the two resistors are connected in series, the same current flows through each. It is now possible to determine the amount of voltage drop across each, Figure WS 9-4.

$$E_1 = I \times R_1$$

$$E_1 = 0.15 \times 300$$

$$E_1 = 45 \text{ volts}$$

$$E_{C(2\text{-}3)} = I \times R_{C(2\text{-}3)}$$

$$E_{C(2\text{-}3)} = 0.15 \times 500$$

$$E_{C(2\text{-}3)} = 75 \text{ volts}$$

Resistor $R_{C(2\text{-}3)}$ is a resistor that was created by combining two resistors connected in parallel with each other. The electrical values of voltage, current, and power associated with this resistor pertain to resistors R_2 and R_3 also. One of the rules concerning parallel circuits states that the voltage across all branches of

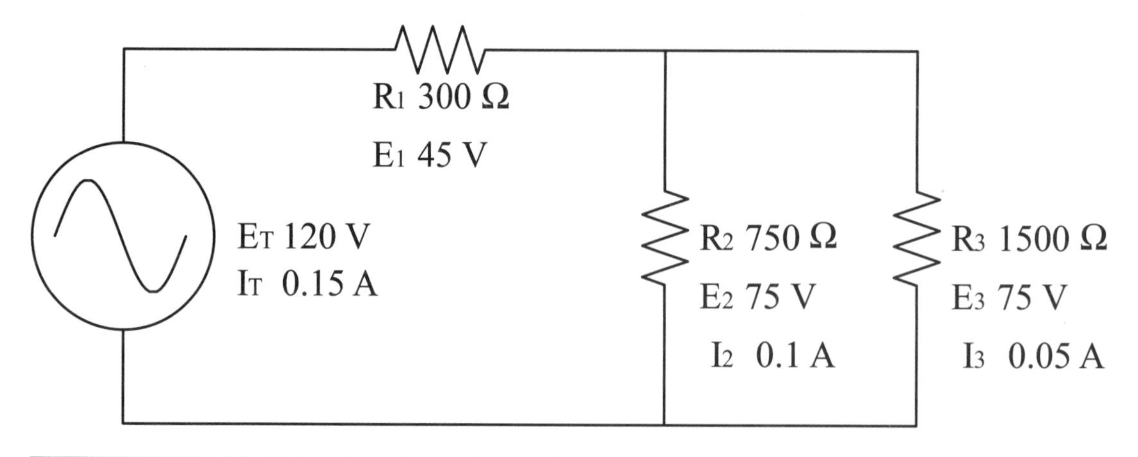

Figure WS 9-5 Determining the remaining values of voltage and current. (Source: Delmar/Cengage Learning)

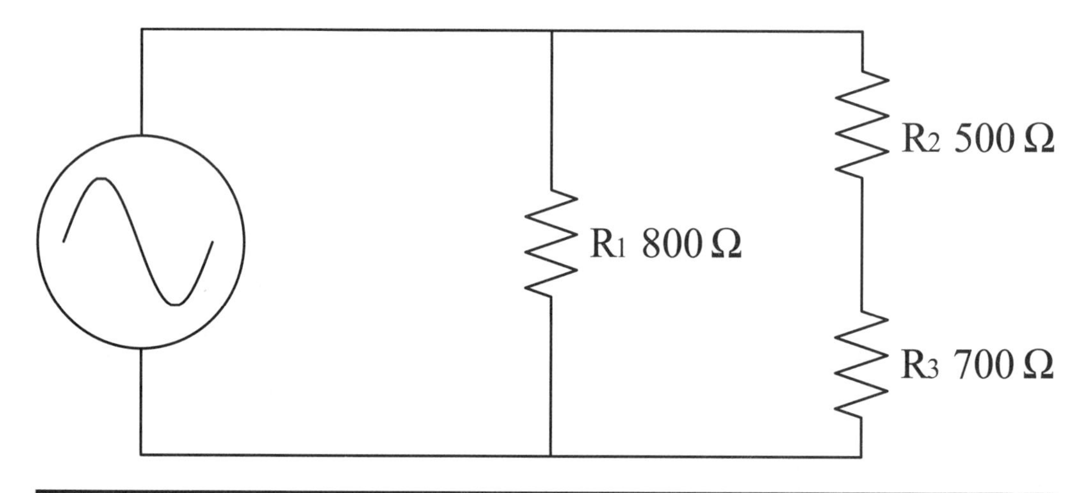

Figure WS 9-6 A second example of a combination circuit. (Source: Delmar/Cengage Learning)

a parallel circuit is the same. Therefore, each of the resistors forming resistor $R_{C(2\text{-}3)}$ will have the same voltage drop as resistor $R_{C(2\text{-}3)}$. Now that the voltage drop across each resistor is known, the current flow through each can be determined, Figure WS 9-5.

$$I_2 = \frac{E_2}{R_2}$$

$$I_2 = \frac{75}{750}$$

$$I_2 = 0.1 \text{ ampere}$$

$$I_2 = \frac{E_2}{R_2}$$

$$I_2 = \frac{75}{1500}$$

$$I_3 = 0.05 \text{ ampere}$$

Another rule concerning parallel circuits is that the total current is the sum of the currents flowing through each branch. In this example, if the currents flowing through resistors R_2 and R_3 are added, they will equal the current flowing through $R_{C(2\text{-}3)}$.

Another example of a simple combination circuit is shown in Figure WS 9-6. To determine which parts are connected in parallel and which are connected in series, again trace the current path through the circuit. Current can leave one side of the

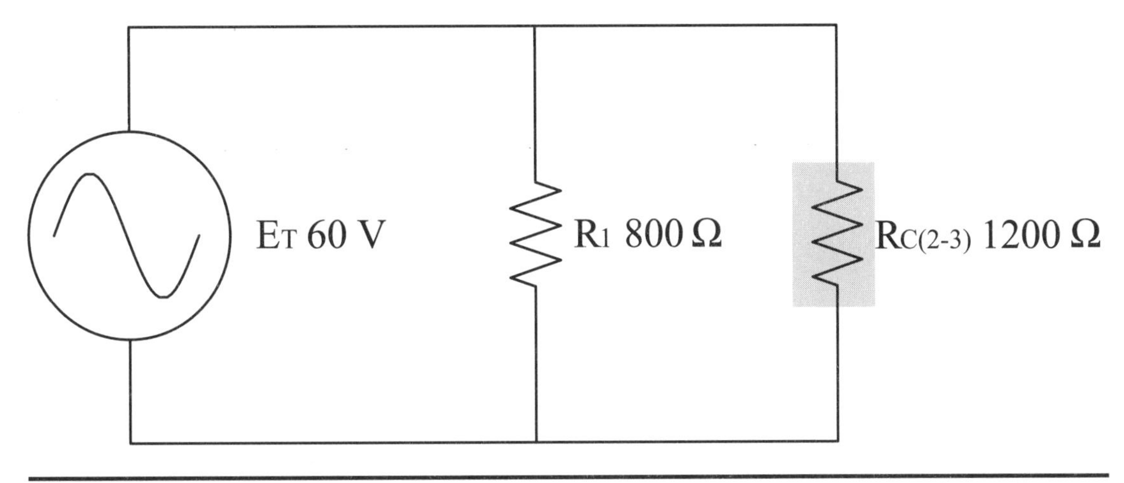

Figure WS 9-7 Resistors R_2 and R_3 are combined to form resistor RC(2-3). (Source: Delmar/Cengage Learning)

alternator and flow through resistor R_1 and return to the other side of the alternator. This provides one current path. A second current path exists through resistors R_2 and R_3. The same current must flow through both resistors R_2 and R_3. Because there is only one current path through resistors R_2 and R_3, they are connected in series with each other. Although resistors R_2 and R_3 are connected in series with each other, they are connected in parallel with resistor R_1.

As in the previous example, the best way to solve values of resistance, voltage, and current for this circuit is to combine resistors to form a simpler circuit, Figure WS 9-7. Because resistors R_2 and R_3 are connected in series with each other, their values can be added to produce a single resistor with a value of 1,200 Ω. The circuit now contains two resistors connected in parallel with each other. The total resistance of the circuit can be determined using the following formula:

$$R_T = \frac{1}{\frac{1}{R_1} + \frac{1}{R_{C(2\text{-}3)}}}$$

$$R_T = \frac{1}{\frac{1}{800} + \frac{1}{1200}}$$

$$R_T = 480\ \Omega$$

In a parallel circuit, the voltage across all branches is the same. Therefore, resistors R_1 and $R_{C(2\text{-}3)}$ have a voltage drop of 60 volts. The amount of current flowing through each resistor can now be determined.

$$I_1 = \frac{E}{R_1}$$

$$I_1 = \frac{60}{800}$$

I_1 = 0.075 ampere or 75 mA

$$I_{C(2\text{-}3)} = \frac{E}{R_{C(2\text{-}3)}}$$

$$I_{C(2\text{-}3)} = \frac{60}{1200}$$

$I_{C(2\text{-}3)}$ = 0.05 ampere or 50 mA

Resistor $R_{C(2\text{-}3)}$ in reality is the combination of resistors R_2 and R_3. The electrical values that pertain to resistor $R_{C(2\text{-}3)}$ also pertain to resistors R_2 and R_3. One of the rules concerning series-connected components is that the current flowing through each component must be the same. Therefore, the current flowing through resistor $R_{C(2\text{-}3)}$ will flow through both resistors R_2 and R_3, Figure WS 9-8. The voltage drop across resistors R_2 and R_3 can now be determined.

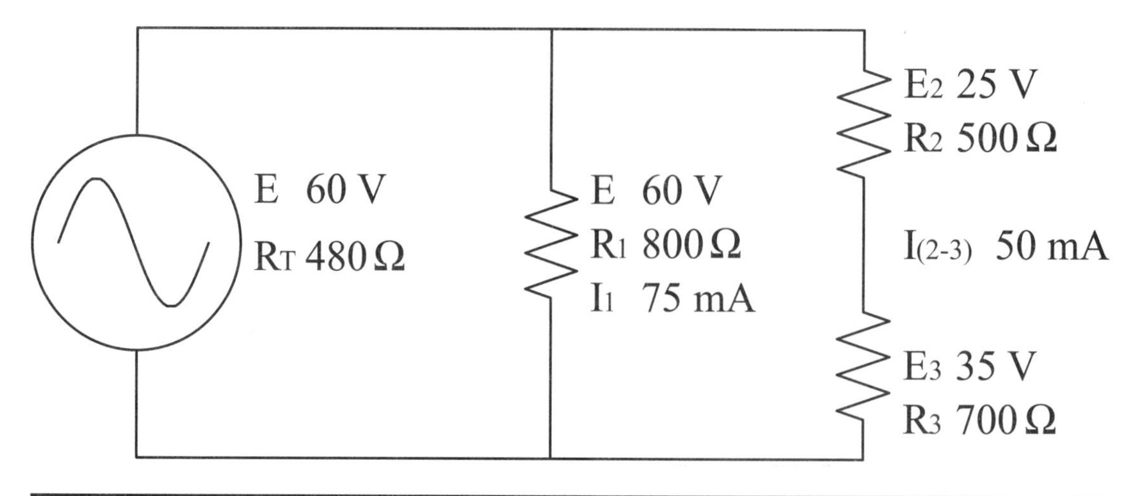

Figure WS 9-8 The current flow is the same through resistors R2 and R3. (Source: Delmar/Cengage Learning)

$$E_2 = 0.05 \times 500$$

$$E_2 = 25 \text{ volts}$$

$$E_3 = 0.05 \times 700$$

$$E_3 = 35 \text{ volts}$$

One of the rules concerning series circuits is that the sum of the voltage drops must equal the applied voltage. The voltage applied to resistors R_2 and R_3 is 60 volts. If the individual voltage drops of 25 volts and 35 volts are added, the sum will be 60 volts.

Determine the indicated values in the following circuits:

1.

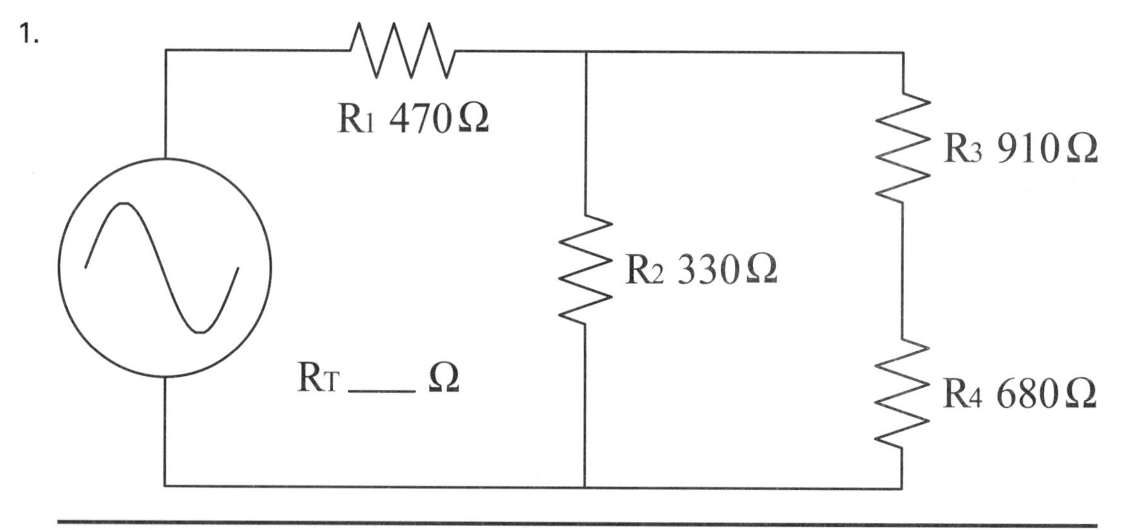

Figure WS 9-9 (Source: Delmar/Cengage Learning)

R_T ____________

2.

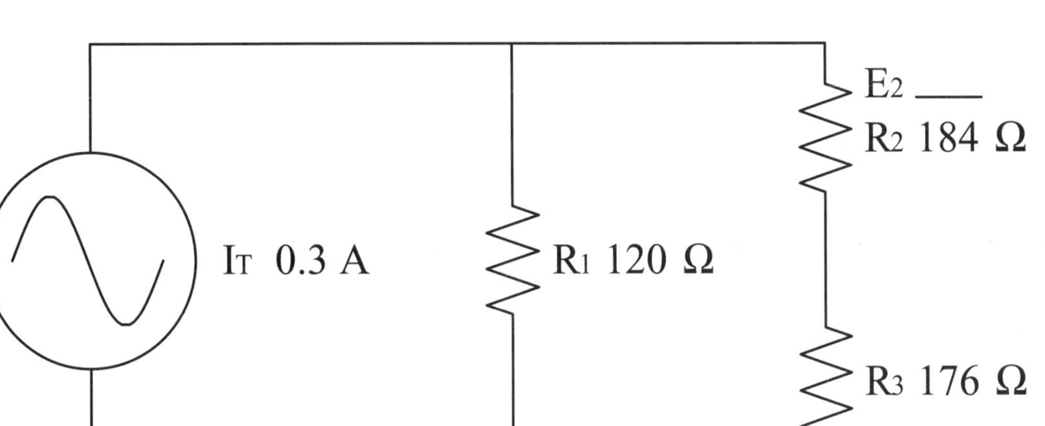

Figure WS 9-10 (Source: Delmar/Cengage Learning)

E_2 ________________

3.

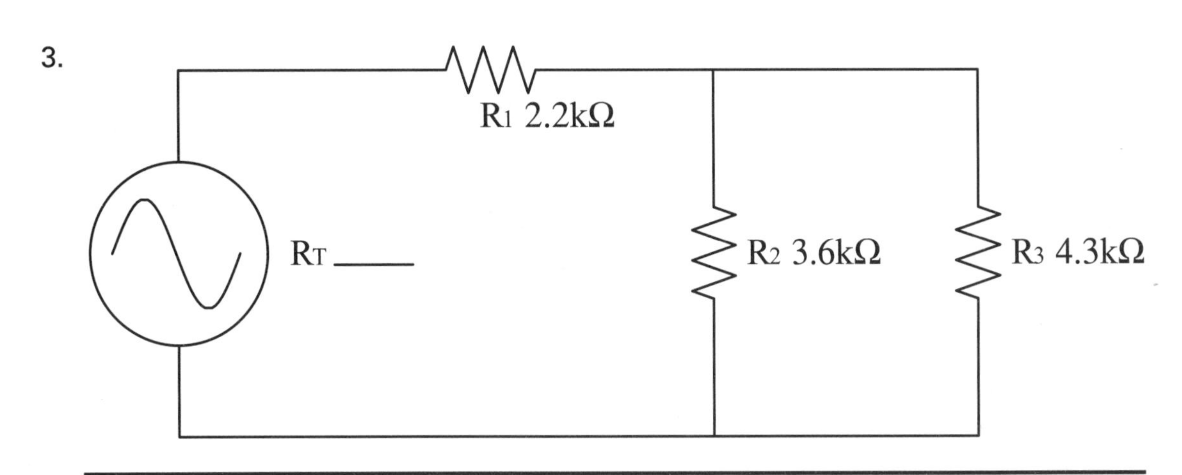

Figure WS 9-11 (Source: Delmar/Cengage Learning)

R_T ________________

4.

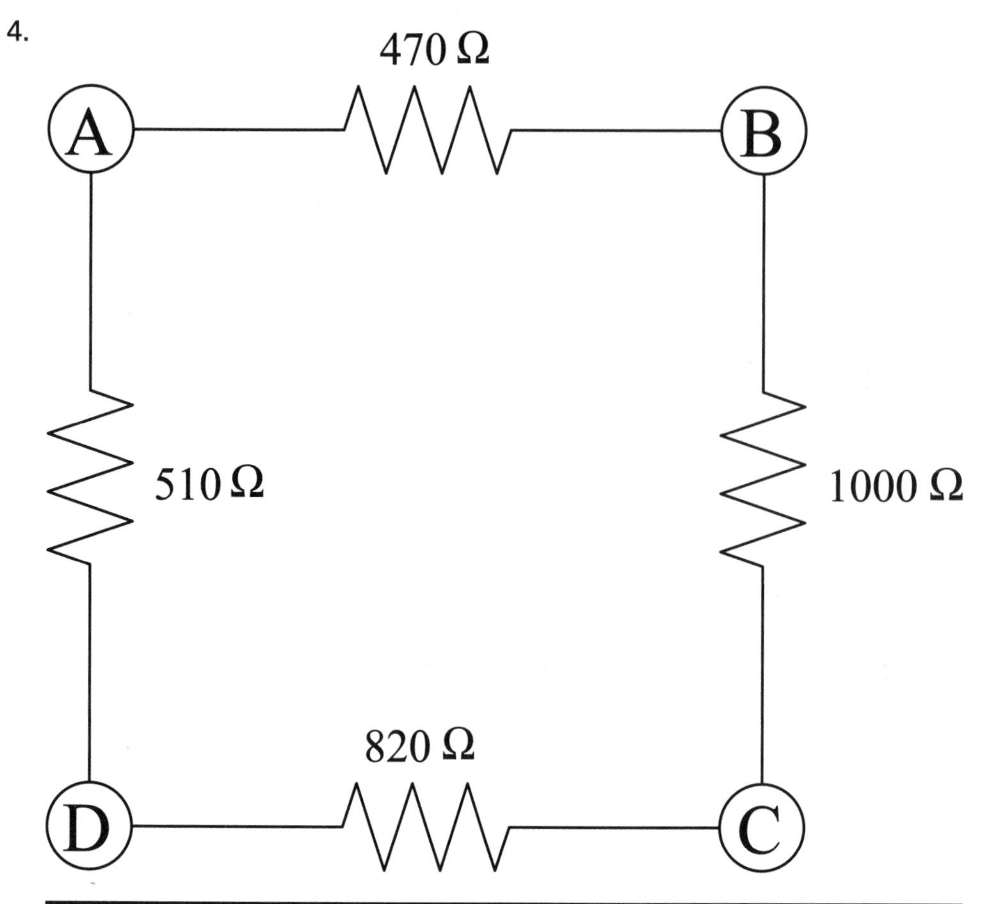

Figure WS 9-12 (Source: Delmar/Cengage Learning)

Determine the total resistance that would be measured by an ohmmeter if it is placed between each of the following points:

A-B ______________

A-C ______________

A-D ______________

B-C ______________

B-D ______________

C-D ______________

5.

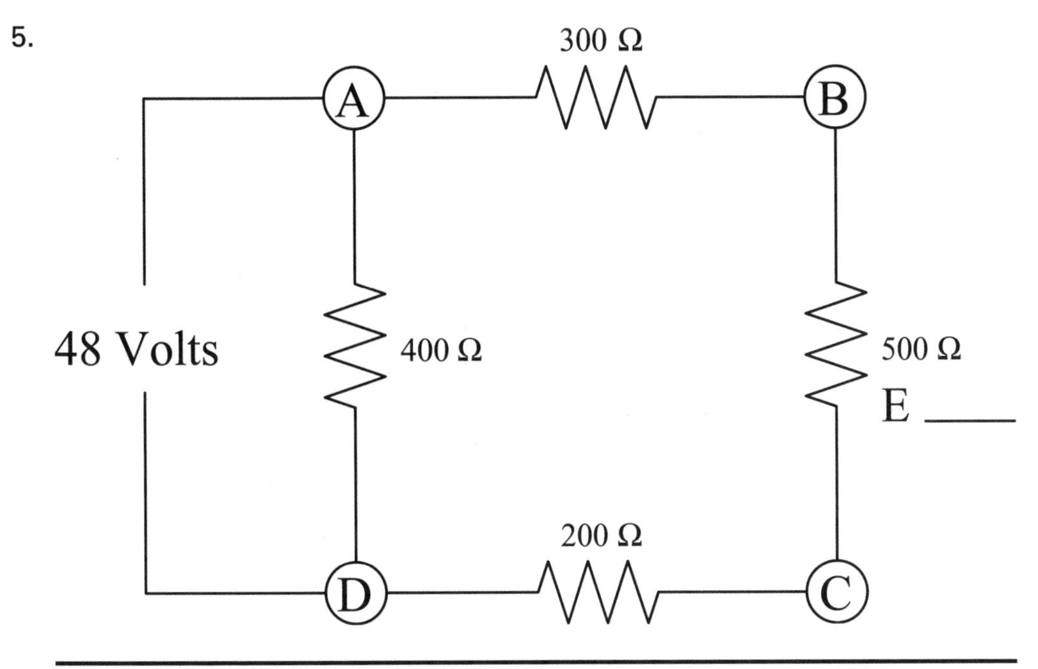

Figure WS 9-13 (Source: Delmar/Cengage Learning)

E ______________

6.

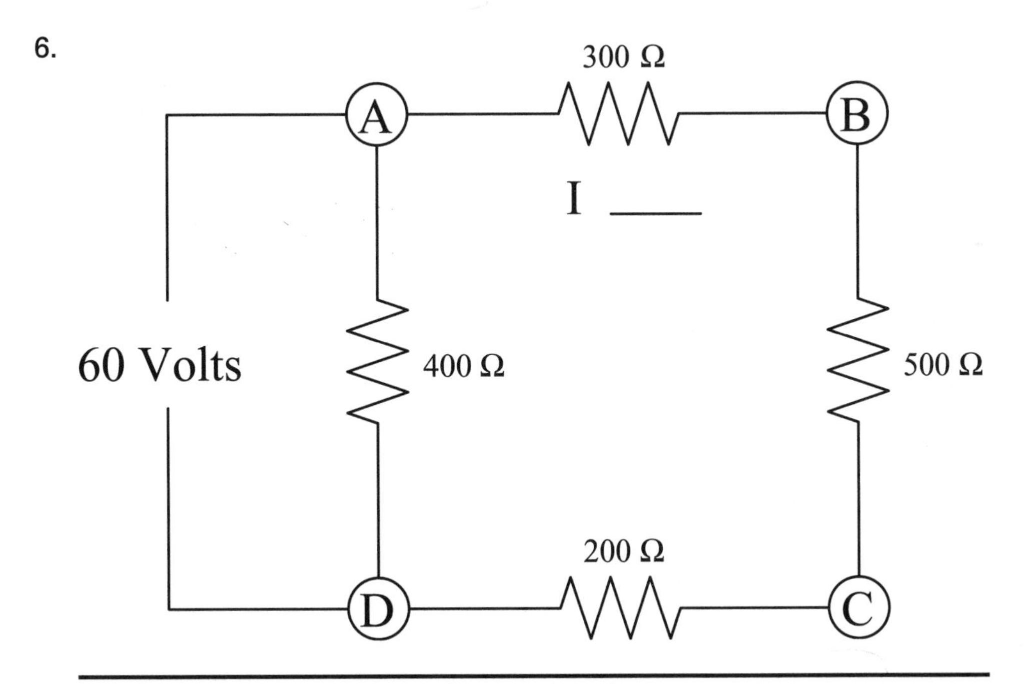

Figure WS 9-14 (Source: Delmar/Cengage Learning)

I ______________

7.

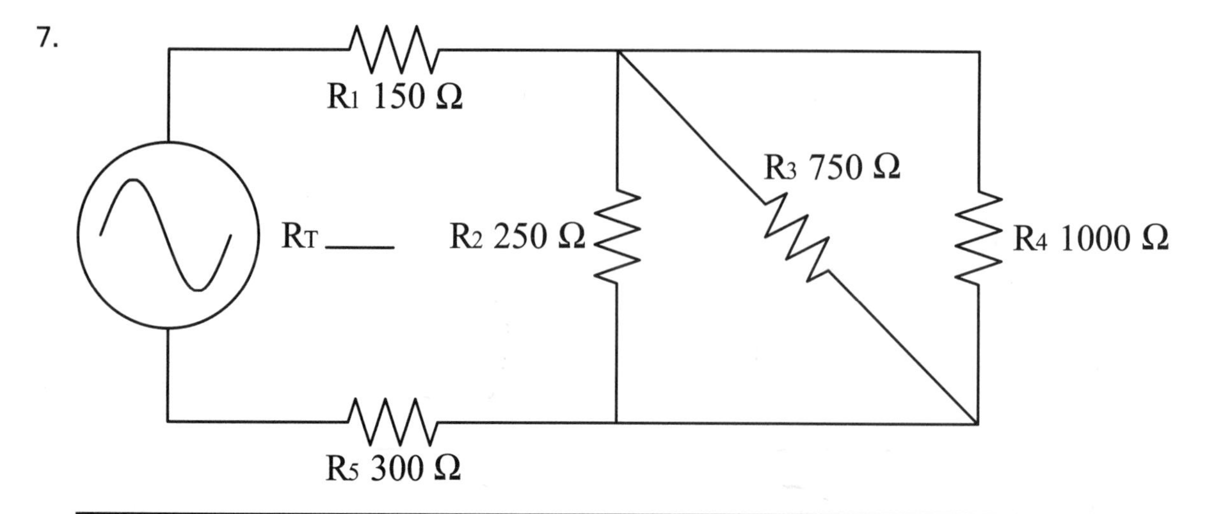

Figure WS 9-15 (Source: Delmar/Cengage Learning)

R_T ______________

8.

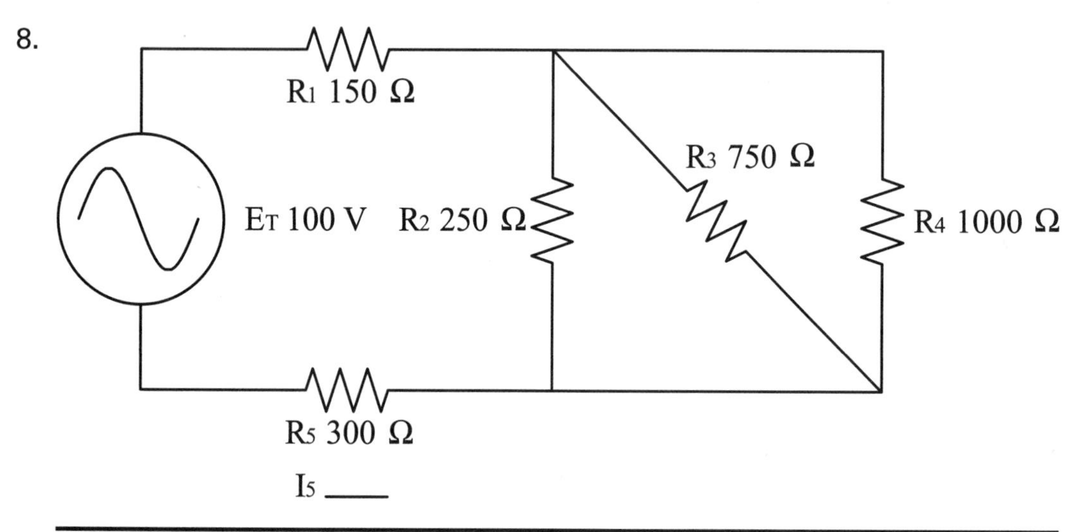

Figure WS 9-16 (Source: Delmar/Cengage Learning)

I_5 ______________

9.

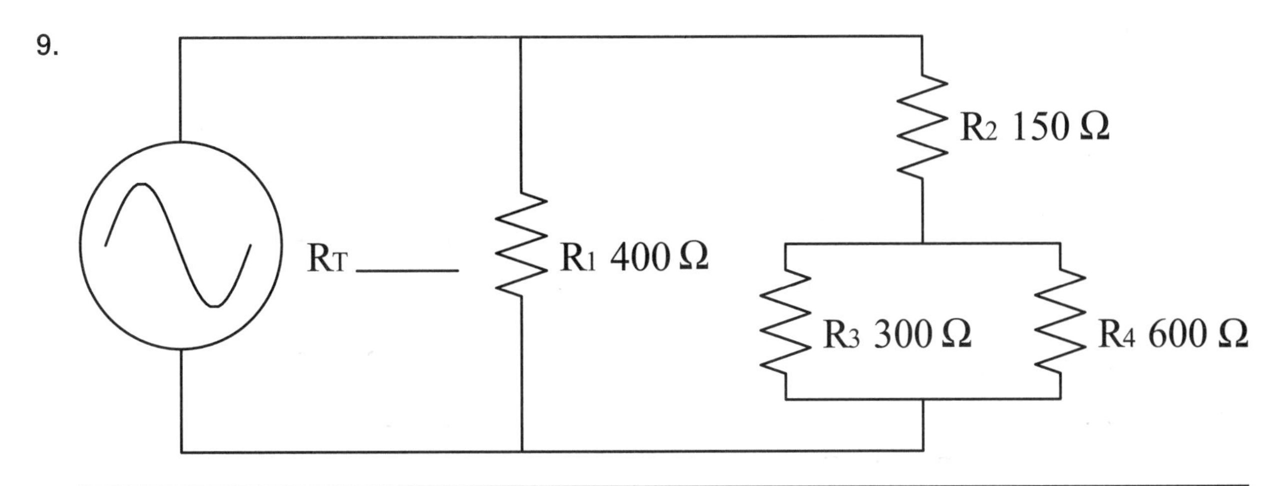

Figure WS 9-17 (Source: Delmar/Cengage Learning)

R_T ____________

10.

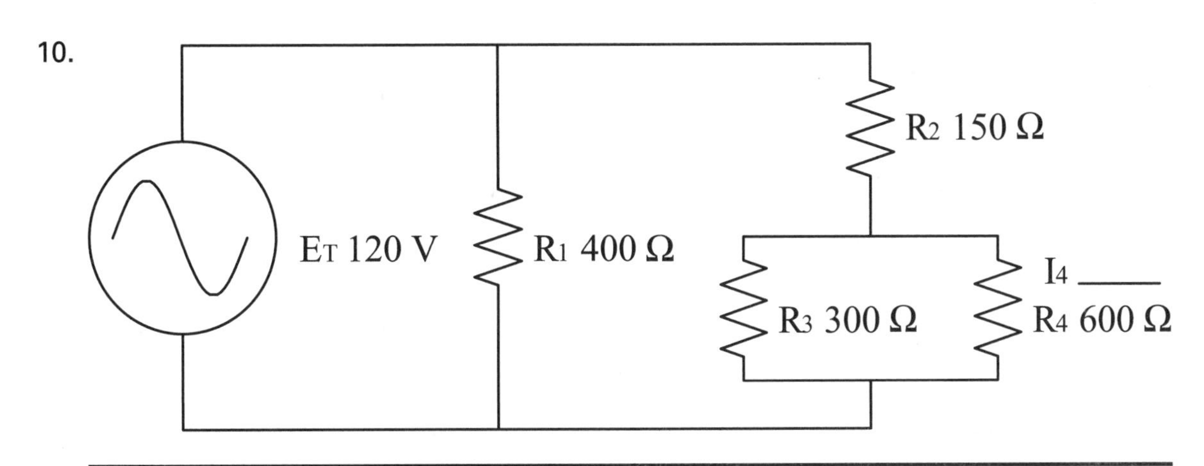

Figure WS 9-18 (Source: Delmar/Cengage Learning)

I_4 ____________

Section 2

LABORATORY EXPERIMENTS

EXPERIMENT 2

Multiple Push-Button Stations

Name:	Date:	Grade:
Comments:		

Objectives

After completing this experiment, the student should be able to:

- Place wire numbers on a schematic diagram
- Place corresponding numbers on control components
- Draw a wiring diagram from a schematic diagram
- Connect a control circuit using two stop and two start push buttons

Materials Needed

- Three-phase power supply
- Three-phase squirrel cage induction motor or simulated load
- Four double acting push buttons (N.O./N.C. on same button)
- Three-phase motor starter or contactor with overload relay containing three load contacts and at least one normally open auxiliary contact
- Control transformer

There may be times when it is desirable to have more than one start-stop push-button station to control a motor. In this experiment the basic start-stop push-button control circuit discussed in Experiment 1 will be modified to include a second stop and start push button.

When a component is used to perform the function of stop in a control circuit, it will generally be a normally closed component and be connected in series with the motor starter coil. In this example, a second stop push button is to be added to an existing start-stop control circuit. The second push-button will be added to the control circuit by connecting it in series with the existing stop push button, Figure 2-1.

When a component is used to perform the function of **start,** it is generally normally open and connected in parallel with the existing start button, Figure 2-2. If either start button is pressed, a circuit will be completed to M coil. When M coil energizes, all M contacts change position. The three load contacts connected between the three-phase power line and the

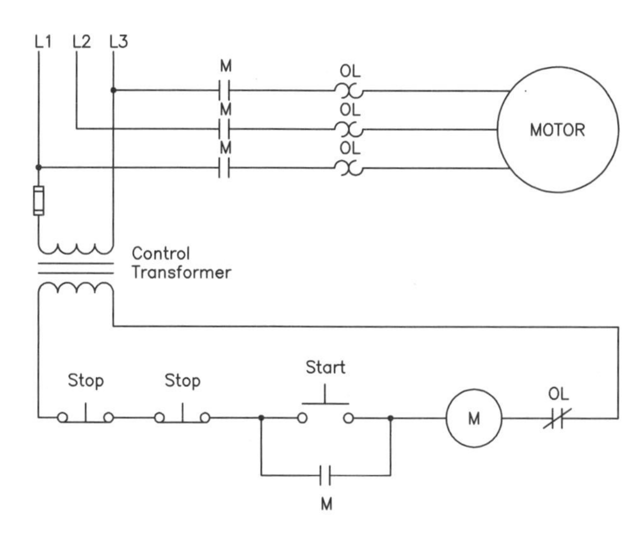

Figure 2-1 Adding a stop button to the circuit

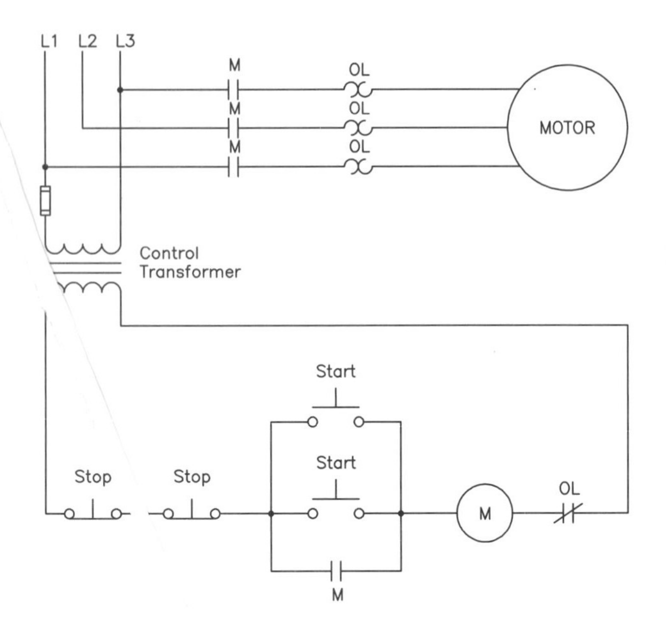

Figure 2-2 A second start button is added to the circuit

motor close to connect the motor to the line. The normally open auxiliary contact connected in parallel with the two start buttons closes to maintain the circuit to M coil when the start button is released.

Developing the Wiring Diagram

Now that the circuit logic has been developed in the form of a schematic diagram, a wiring diagram will be drawn from the schematic. The components needed to connect this circuit are shown in Figure 2-3. Following the same procedure discussed in Experiment 1, wire numbers will be placed on the schematic diagram, Figure 2-4. After wire numbers are placed on the schematic, corresponding numbers will be placed on the control components, Figure 2-5.

Connecting the Circuit

1. Using the schematic in Figure 2-4 or the diagram with numbered components in Figure 2-5, connect the circuit in the laboratory by connecting all like numbers together.
2. After the circuit has been connected, check with your instructor before turning on the power.

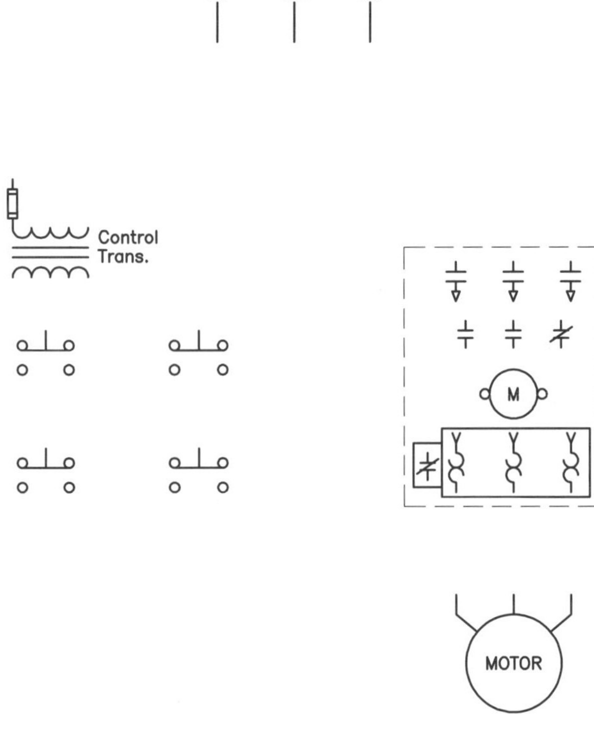

Figure 2-3 Components needed to produce a wiring diagram

3. Turn on the power and test the circuit for proper operation.
4. Turn off the power and disconnect the circuit. Return all components to their proper place.

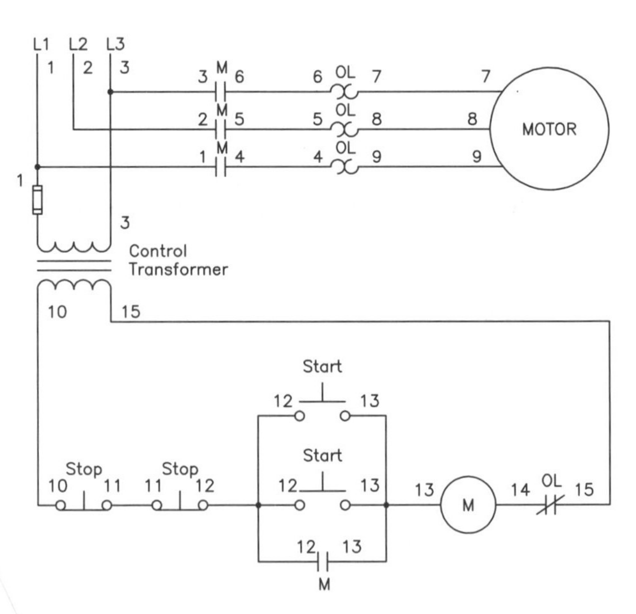

Figure 2-4 Numbering the schematic diagram

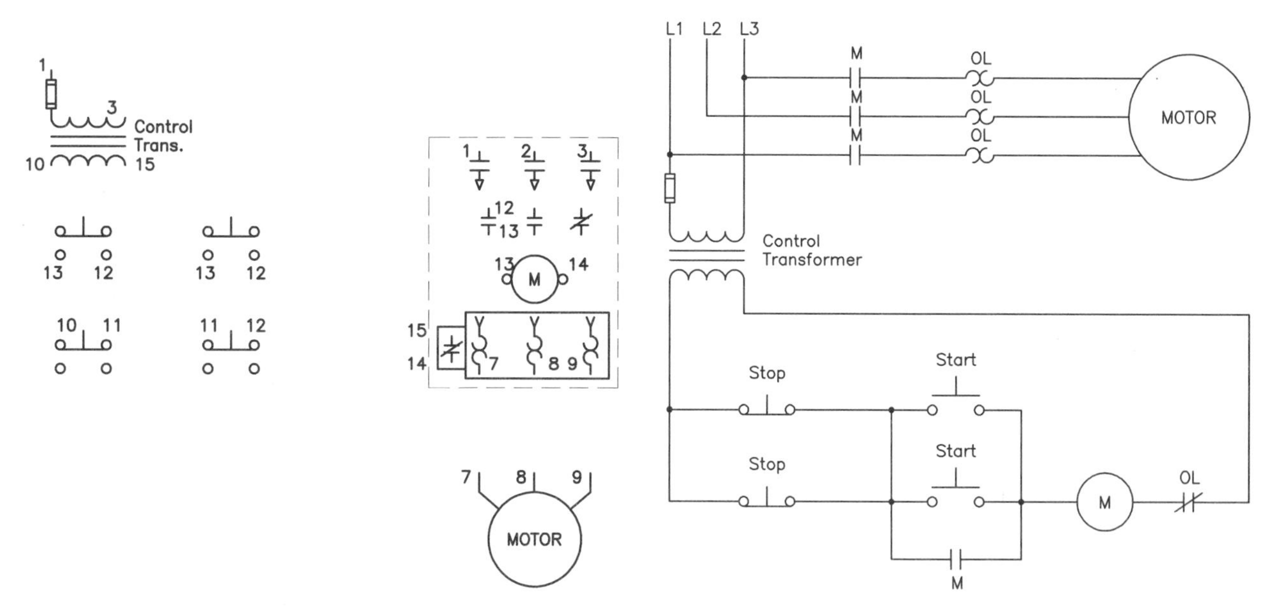

Figure 2-5 Numbering the components

Figure 2-6 The stop buttons are connected in parallel

Review Questions

1. When a component is to be used for the function of start, is the component generally normally open or normally closed?

2. When a component is to be used for the function of stop, is the component generally normally open or normally closed?

3. The two stop push buttons in Figure 2-2 are connected in series with each other. What would be the action of the circuit if they were to be connected in parallel, as shown in Figure 2-6?

4. What would be the action of the circuit if both start buttons were to be connected in series, as shown in Figure 2-7?

5. Following the procedure discussed in Experiment 1, place wire numbers on the schematic in Figure 2-7. Place corresponding wire numbers on the components shown in Figure 2-8.

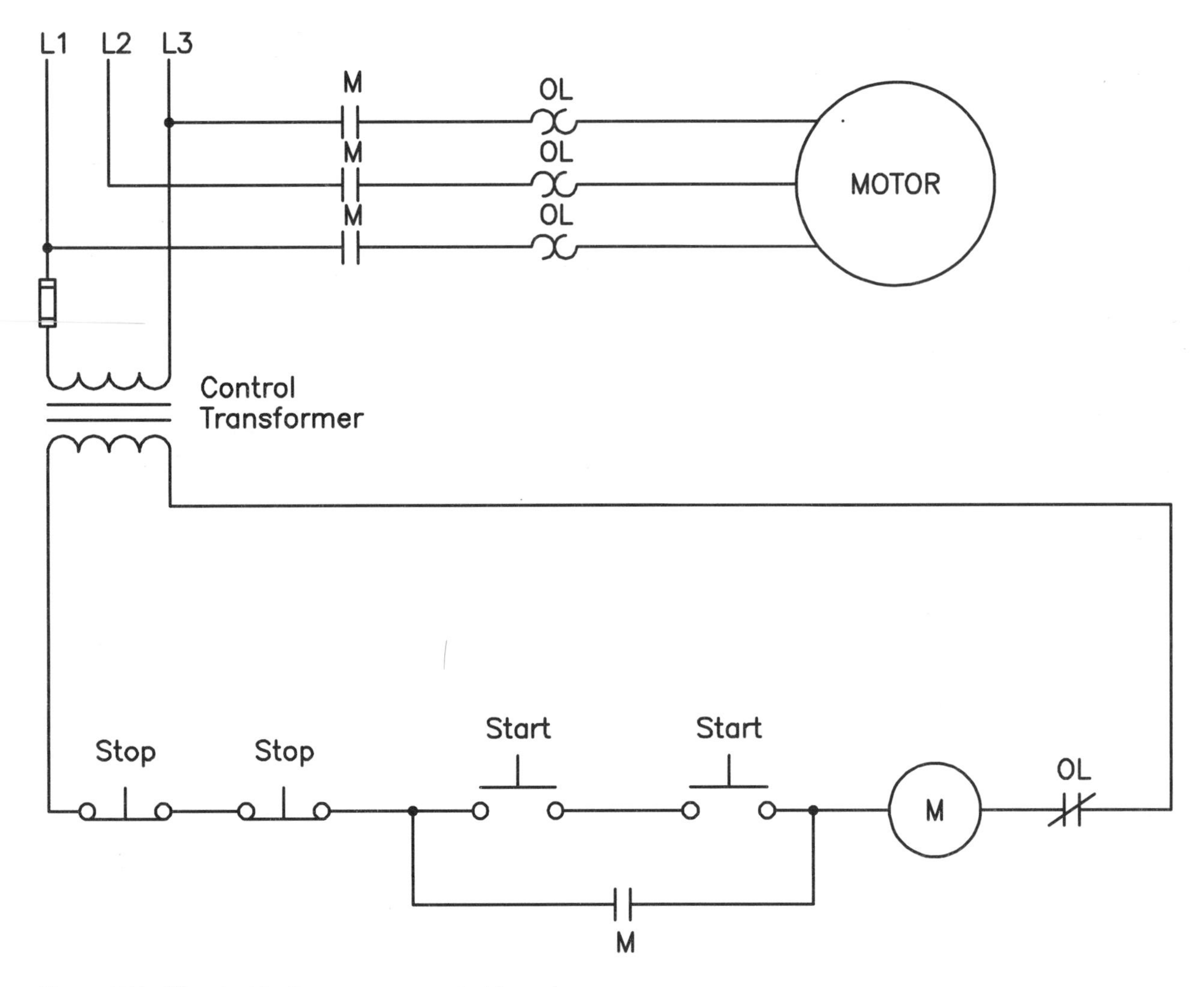

Figure 2-7 The start buttons are connected in series

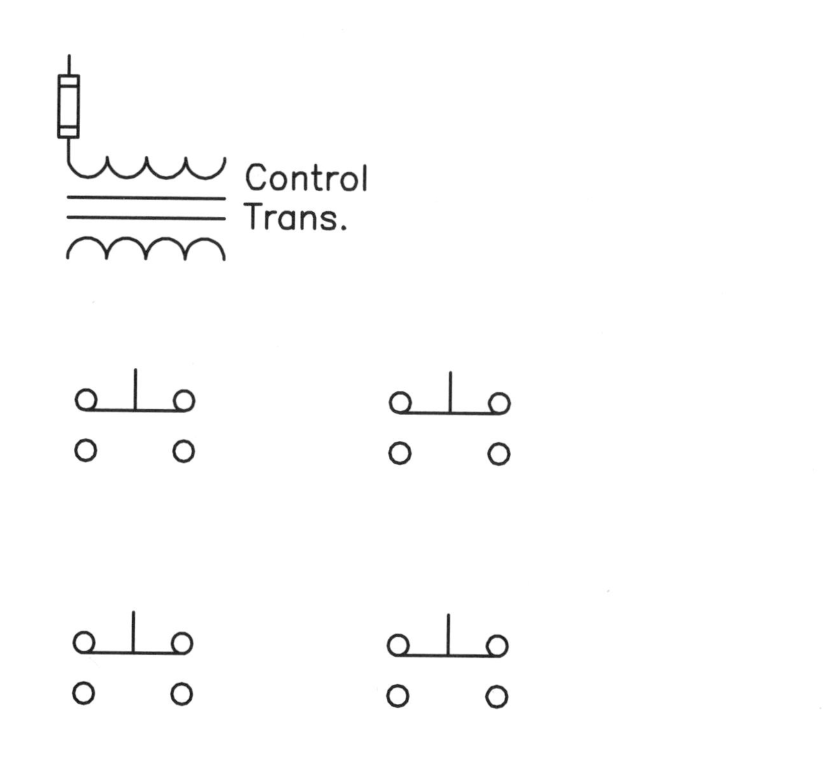

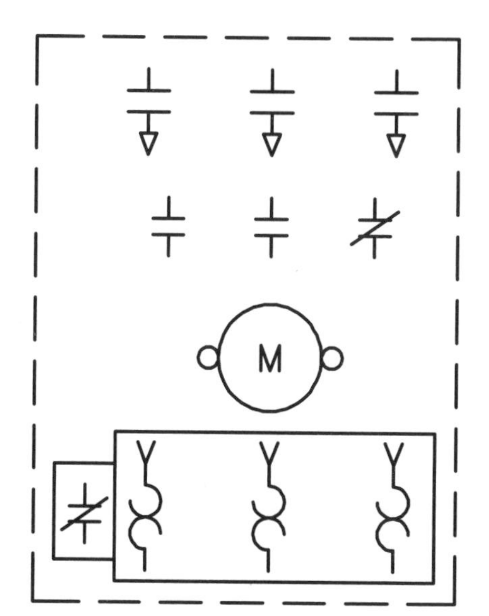

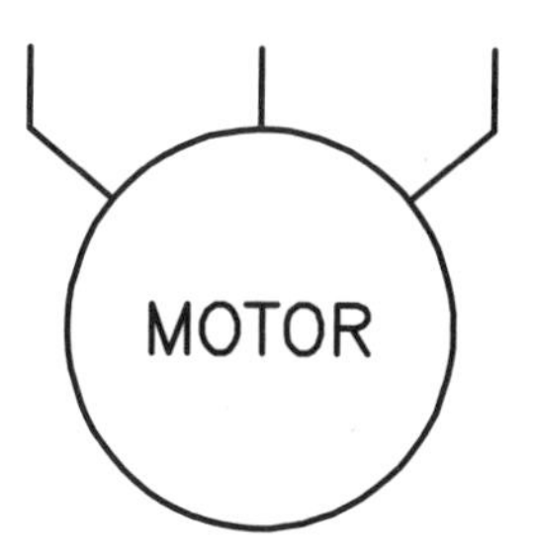

Figure 2-8 Add wire numbers to these components

EXPERIMENT 3

Forward-Reverse Control

Name:	Date:	Grade:
Comments:		

Objectives

After completing this experiment, the student should be able to:

- Discuss cautions that must be observed in reversing circuits
- Explain how to reverse a three-phase motor
- Discuss interlocking methods
- Connect a forward-reverse motor control circuit

Materials Needed

- Three-phase power supply
- Control transformer
- One of the following:
  1. A three-phase reversing starter
  2. Two three-phase contactors with at least one normally open and one normally closed auxiliary contact on each contactor; one three-phase overload relay or three single-phase overload relays
- Three-phase squirrel cage motor or simulated motor load
- Three double acting push buttons (N.O./N.C. on each button)

The direction of rotation of any three-phase motor can be reversed by changing any two motor T leads. Because the motor is connected to the power line regardless of which direction it operates, a separate contactor is needed for each direction. Because only one motor is in operation, however, only one overload relay is needed to protect the motor. True reversing controllers contain two separate contactors and one overload relay built into one unit. (Refer to Figures 39-1A and 39-3A in *Industrial Motor Control,* 4e.)

Interlocking

Interlocking prevents some action from taking place until some other action has been performed. In the case of reversing starters, interlocking is used to prevent both contactors from being energized at the same time. This would result in two of the three phase lines being shorted together. Interlocking forces one contactor to be de-energized before the other one can be energized.

Most reversing controllers contain mechanical interlocks as well as electrical interlocks. Mechanical interlocking is accomplished by using the contactors to operate a mechanical lever that prevents the other contactor from closing while the other is energized.

Electrical interlocking is accomplished by connecting the normally closed auxiliary contacts on one contactor in series with the coil of the other contactor, Figure 3-1. Assume that the forward push button is pressed and F coil energizes. This causes all F contacts to change position. The three F load contacts close and connect the motor to the line. The normally open F auxiliary contact closes to maintain the circuit when the forward push button is released, and the normally closed F auxiliary contact connected in series with R coil opens, Figure 3-2. (***Note:*** Figure 3-2 illustrates the circuit as it is when the forward starter has been energized. Schematics of this type are used throughout this laboratory manual to help students understand how relay logic operates. This can lead to confusion, however, because contacts that are connected normally open will be shown closed and normally closed contacts will be shown open. To help avoid confusion, normally open contacts that are closed during the stage the circuit is in at that moment will use double lines to indicate that the contact is now closed. Contacts that are normally closed but open at that stage of circuit operation will show a line at the edges of the contact, but the contact will be open in the middle, Figure 3-3.)

If the opposite direction of rotation is desired, the stop button must be pressed first. If the reverse push button were to be pressed first, the now open F auxiliary contact connected in series with R coil would prevent a complete circuit from being established. After the stop button has been pressed, however, F coil de-energizes and all F contacts return to their normal position. The reverse push button can now be pressed to energize R coil, Figure 3-4. When R coil energizes, all R contacts change position. The three R load contacts close and connect the motor to the line. Notice, however, that two of the motor T leads are connected to different lines. The normally closed R auxiliary contact opens to prevent the possibility of F coil being energized until R coil is de-energized.

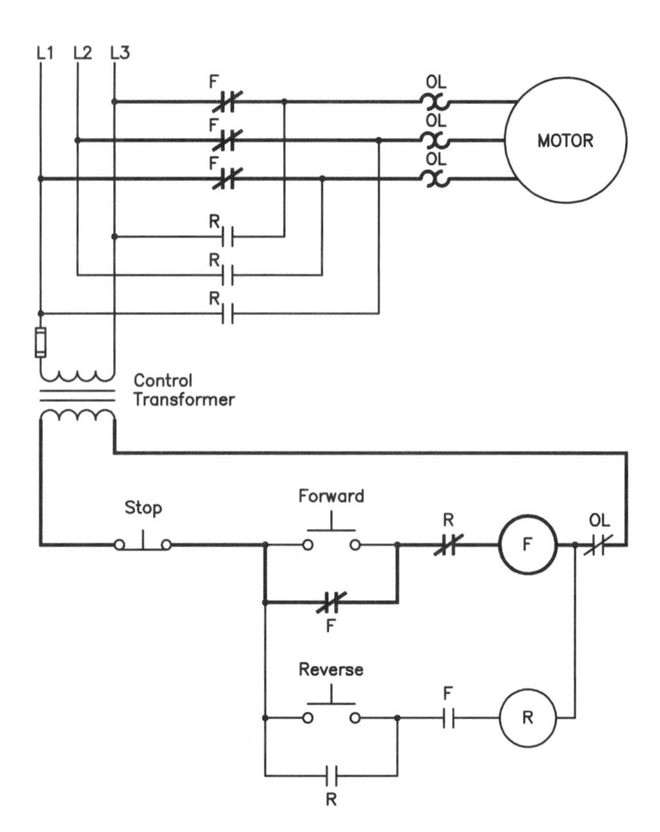

Figure 3-2 Motor operating in the forward direction

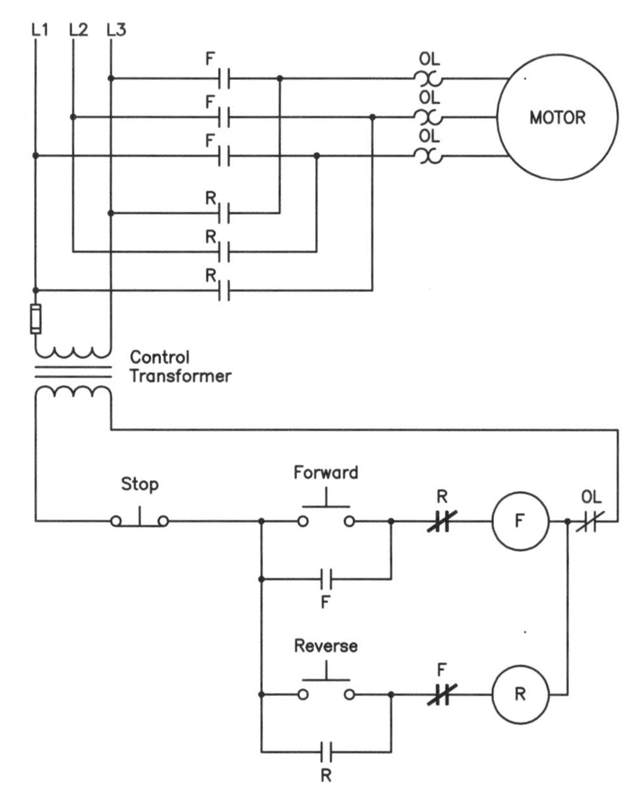

Figure 3-1 Forward-reverse control with interlock

Figure 3-3 Contacts to illustrate circuit logic

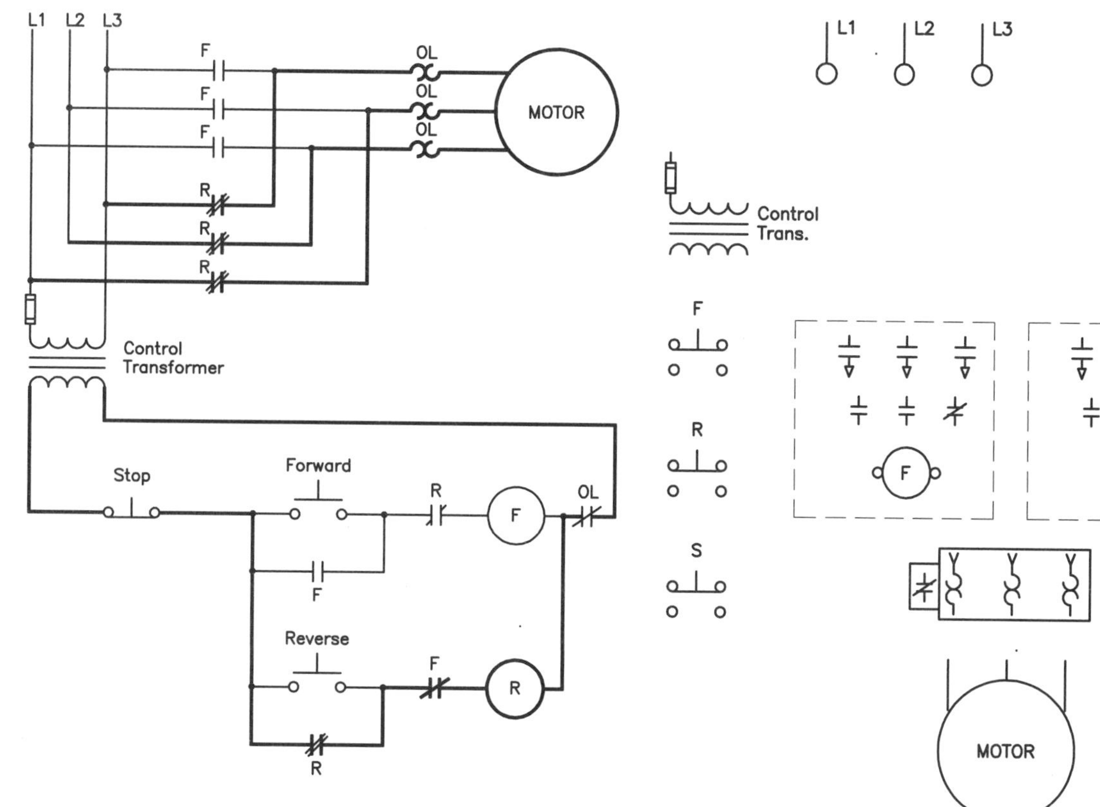

Figure 3-4 Motor operating in the reverse direction

Figure 3-5 Components needed to construct a reversing control circuit

Developing a Wiring Diagram

The same basic procedure will be used to develop a wiring diagram from the schematic as was followed in the previous experiments. The components needed to construct this circuit are shown in Figure 3-5. In this example it is assumed that two contactors and a separate three-phase overload relay are being used.

The first step is to place wire numbers on the schematic diagram. A suggested numbering sequence is shown in Figure 3-6. The next step is to place the wire numbers beside the corresponding components of the wiring diagram, Figure 3-7.

Wiring the Circuit

1. Using the components listed at the beginning of this experiment, connect a forward-reverse control circuit with interlocks. Connect the control section of the circuit before connecting the load section. This will permit the control circuit to be tested without the possibility of

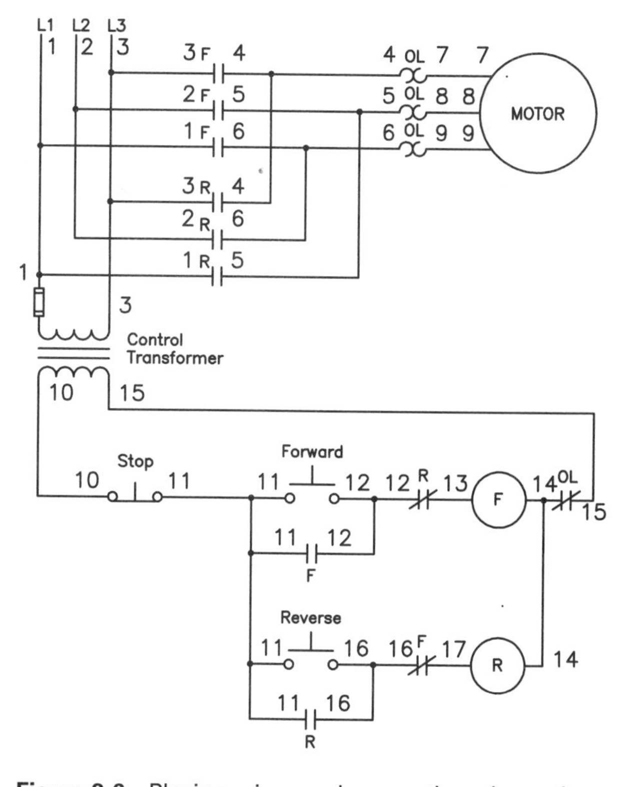

Figure 3-6 Placing wire numbers on the schematic

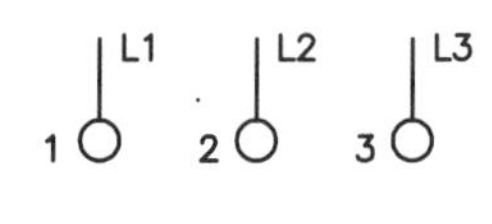

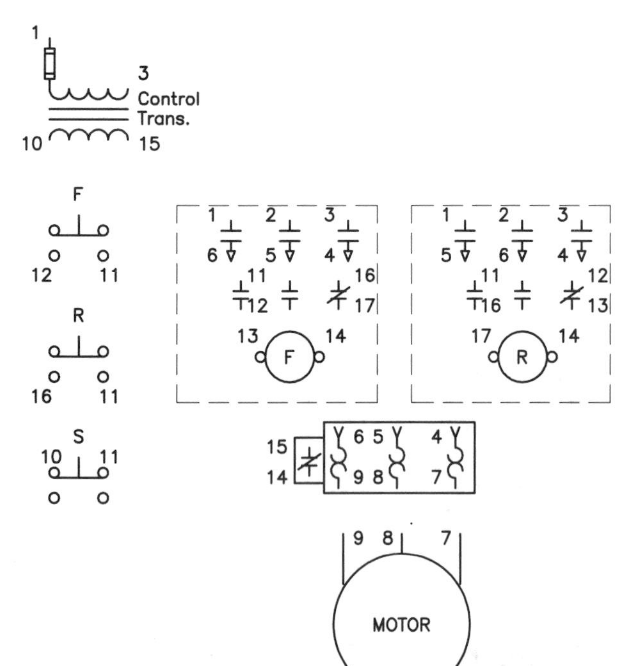

Figure 3-7 Placing corresponding wire numbers on the components

shorting two of the three phase lines together.

2. After checking with the instructor, turn on the power and test the control section of the circuit for proper operation.
3. Turn off the power and complete the wiring by connecting the load portion of the circuit.
4. Turn on the power and test the motor for proper operation.
5. Turn off the power and disconnect the circuit. Return the components to their proper place.

Review Questions

1. How can the direction of rotation of a three-phase motor be changed?

2. What is interlocking?

3. Referring to the schematic shown in Figure 3-1, how would the circuit operate if the normally closed R contact connected in series with F coil were to be connected normally open?

4. What would be the danger, if any, if the circuit were to be wired as stated in review question 3?

5. How would the circuit operate if the normally closed auxiliary contacts were to be connected so that F contact was connected in series with F coil and R contact was connected in series with R coil, Figure 3-8?

6. Assume that the circuit shown in Figure 3-1 were to be connected as shown in Figure 3-9. In what way would the operation of the circuit be different, if at all?

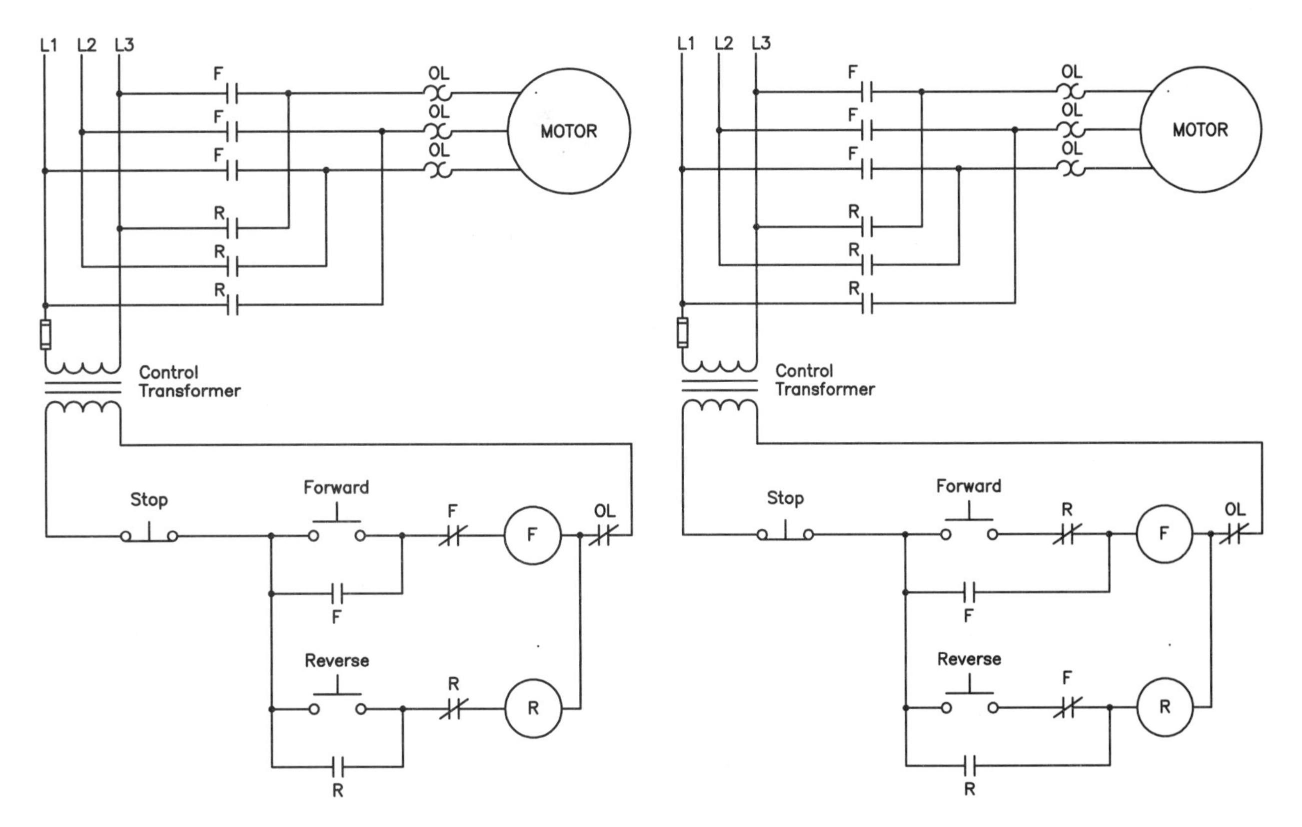

Figure 3-8 F and R normally open auxiliary contacts are connected incorrectly

Figure 3-9 The position of the holding contacts has been changed

EXPERIMENT 4

Sequence Control

Name: Date: Grade:

Comments:

Objectives

After completing this experiment, the student should be able to:

- Define sequence control
- Discuss methods of obtaining sequence control
- Connect a control circuit for three motors that must be started in a predetermined sequence

Materials Needed

- Three-phase power supply
- Control transformer
- Three motor starters containing at least three load contacts and two normally open auxiliary contacts
- Three squirrel cage motors or three simulated motor loads
- Four double acting push buttons (N.O./N.C. on each button)

Sequence control forces a circuit to operate in a predetermined manner. In this experiment three motors are to be started in sequence from 1 to 3. The requirements for the circuit are as follows:

1. The motors must start in sequence from 1 to 3. For example, motor 1 must be started before motor 2 can be started, and motor 2 must start before motor 3 can be started. Motor 2 cannot start before motor 1, and motor 3 cannot start before motor 2.
2. Each motor is started by a separate push button.
3. One stop button will stop all motors.
4. An overload on any motor will stop all three motors.

As a general rule, there is more than one way to design a circuit that will meet the specified requirements, just as there is generally more than one road that can be taken to reach a destination. One design that will meet the requirements is shown in Figure 4-1. Because the logic of the circuit is of primary interest, the load contacts and motors are not shown. In this circuit, push button 1 must be pressed before power can be provided to push button 2. When motor starter 1 energizes, the normally open auxiliary contact 1M closes, providing power to coil 1M and to push button 2. Motor starter 2 can now be started by pressing push button 2. After motor starter 2 energizes, auxiliary contact 2M closes and provides power to coil 2M and push button 3. If the stop button should be pressed or any overload contact open, power will be interrupted to all starters.

A Second Circuit for Sequence Control

A second method of providing sequence control is shown in Figure 4-2. In this circuit,

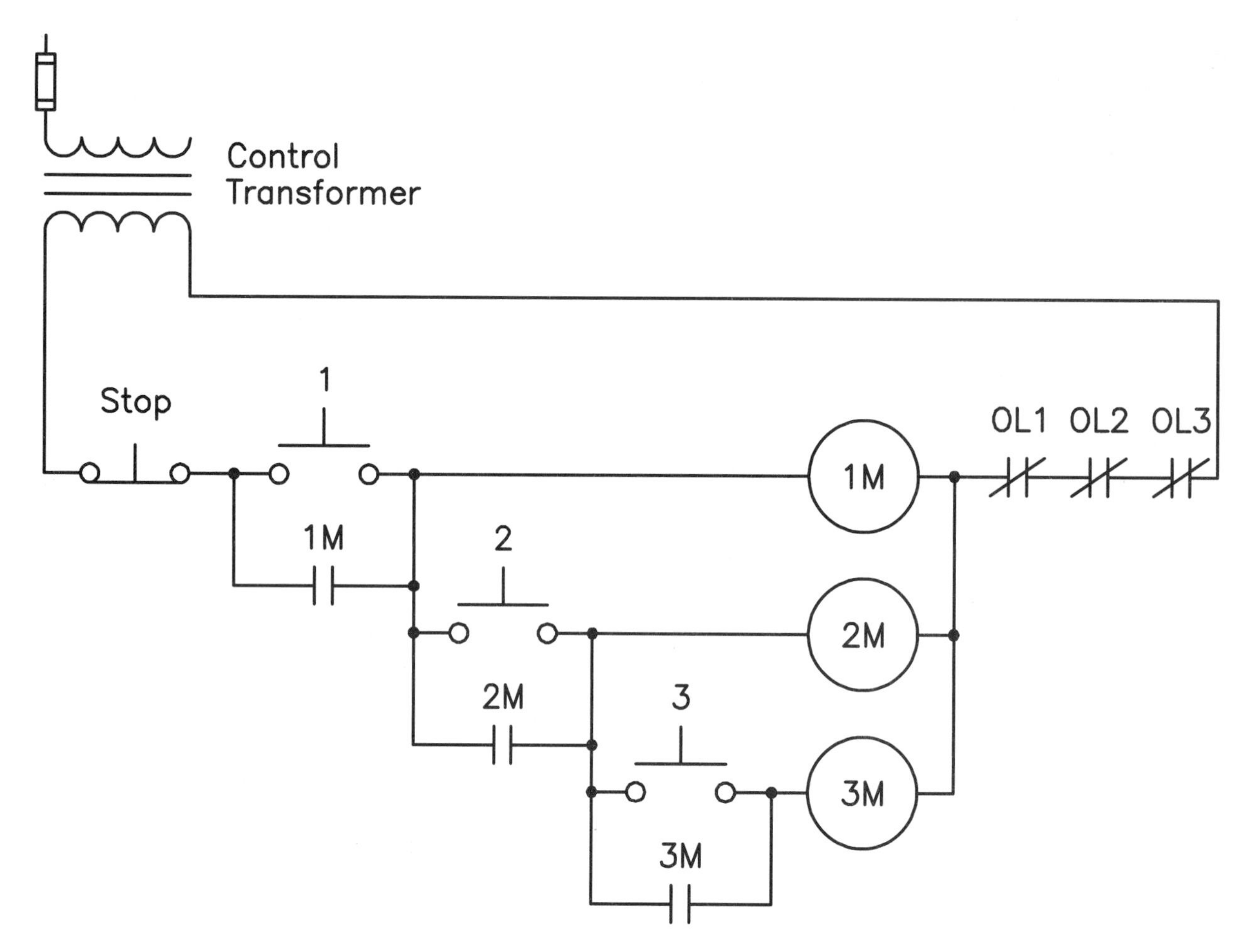

Figure 4-1 First example of starting three motors in sequence

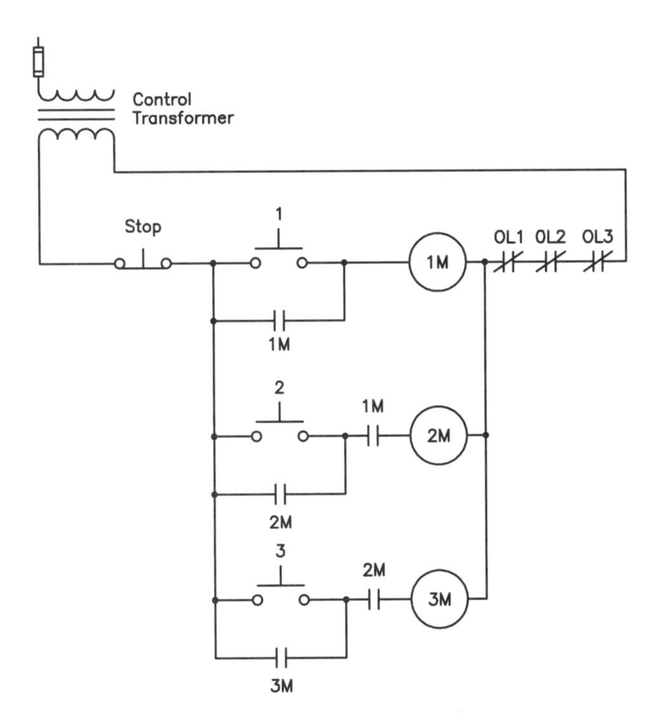

Figure 4-2 A second circuit for sequence control

normally open auxiliary contacts located on motor starters 1M and 2M are used to ensure that the three motors start in the proper sequence. A normally open 1M auxiliary contact connected in series with starter coil 2M prevents motor 2 from starting before motor 1, and a normally open 2M auxiliary contact connected in series with coil 3M prevents motor 3 from starting before motor 2. If the stop button should be pressed or if any overload contact should open, power will be interrupted to all starters.

Developing a Wiring Diagram

The schematic shown in Figure 4-2 is shown with the motors in Figure 4-3. A drawing of the components needed to connect this circuit is shown in Figure 4-4. The schematic diagram shown in Figure 4-3 is shown with wire numbers in Figure 4-5. The components with corresponding wire numbers are shown in Figure 4-6.

Connecting the Circuit

1. Using the materials listed at the beginning of this experiment, connect the circuit shown in Figure 4-5. Follow the number sequence shown.
2. After checking with the instructor, turn on the power and test the circuit for proper operation.
3. Turn off the power and disconnect the circuit.
4. Using the schematic diagram shown in Figure 4-1, add wire numbers to the schematic.
5. Place these wire numbers beside the proper components shown in Figure 4-4.
6. Connect the circuit shown in Figure 4-1 by following the wire numbers placed on the schematic.
7. After checking with the instructor, turn on the power and test the circuit for proper operation.
8. Turn off the power and disconnect the circuit. Return the components to their proper place.

Review Questions

1. What is the purpose of sequence control?

2. Refer to the schematic diagram in Figure 4-5. Assume that the 1M contact located between wire numbers 29 and 30 had been connected normally closed instead of normally open. How would this circuit operate?

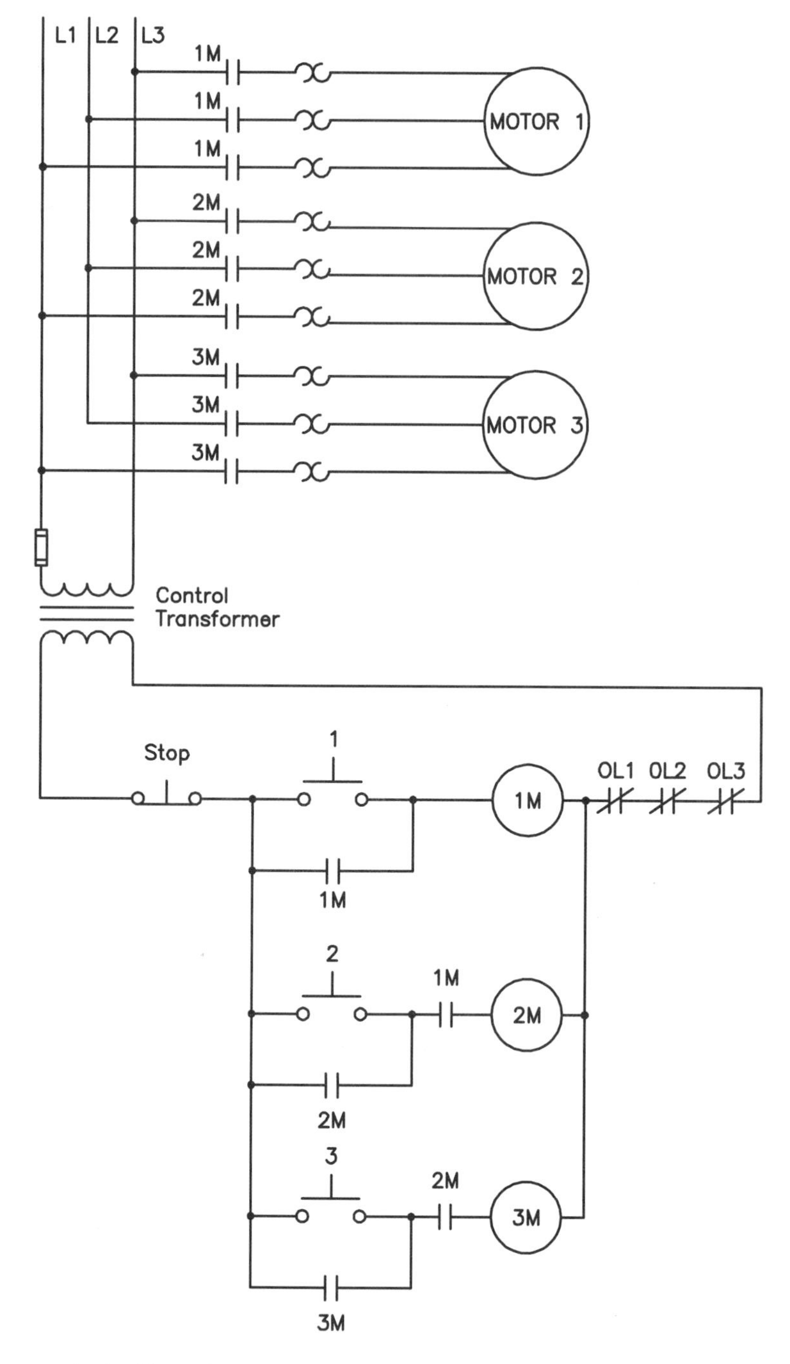

Figure 4-3 Sequence control with motors

3. Assume that all three motors shown in Figure 4-5 are running. Now assume that the stop button is pressed and motors 1 and 2 stop running, but motor 3 continues to operate. Which of the following could cause this problem?
   a. Stop button is shorted.
   b. 2M contact between wire numbers 31 and 32 is hung closed.
   c. The 3M load contacts are welded shut.
   d. The normally open 3M contact between wire numbers 23 and 31 is hung closed.

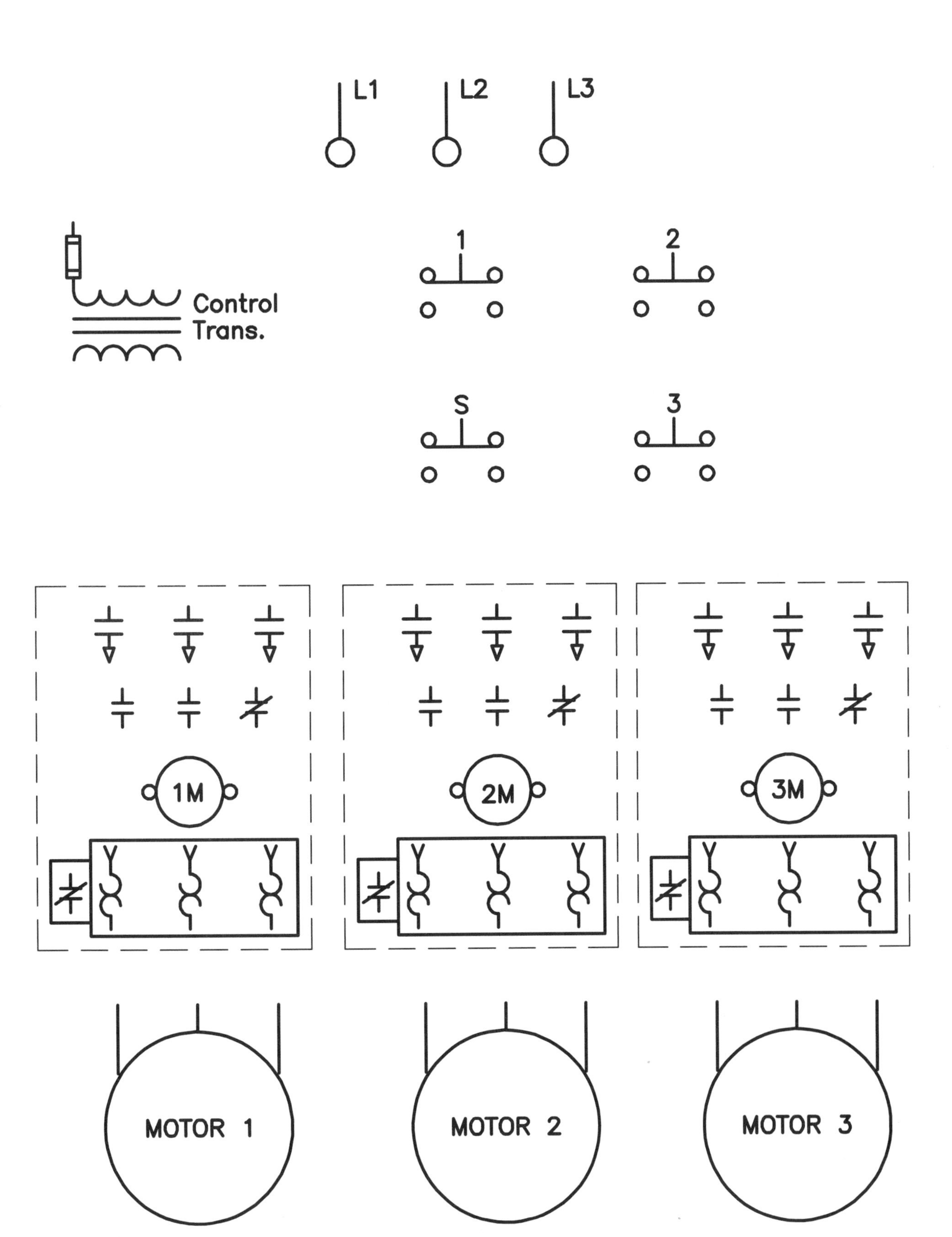

Figure 4-4 Components needed to connect the circuit

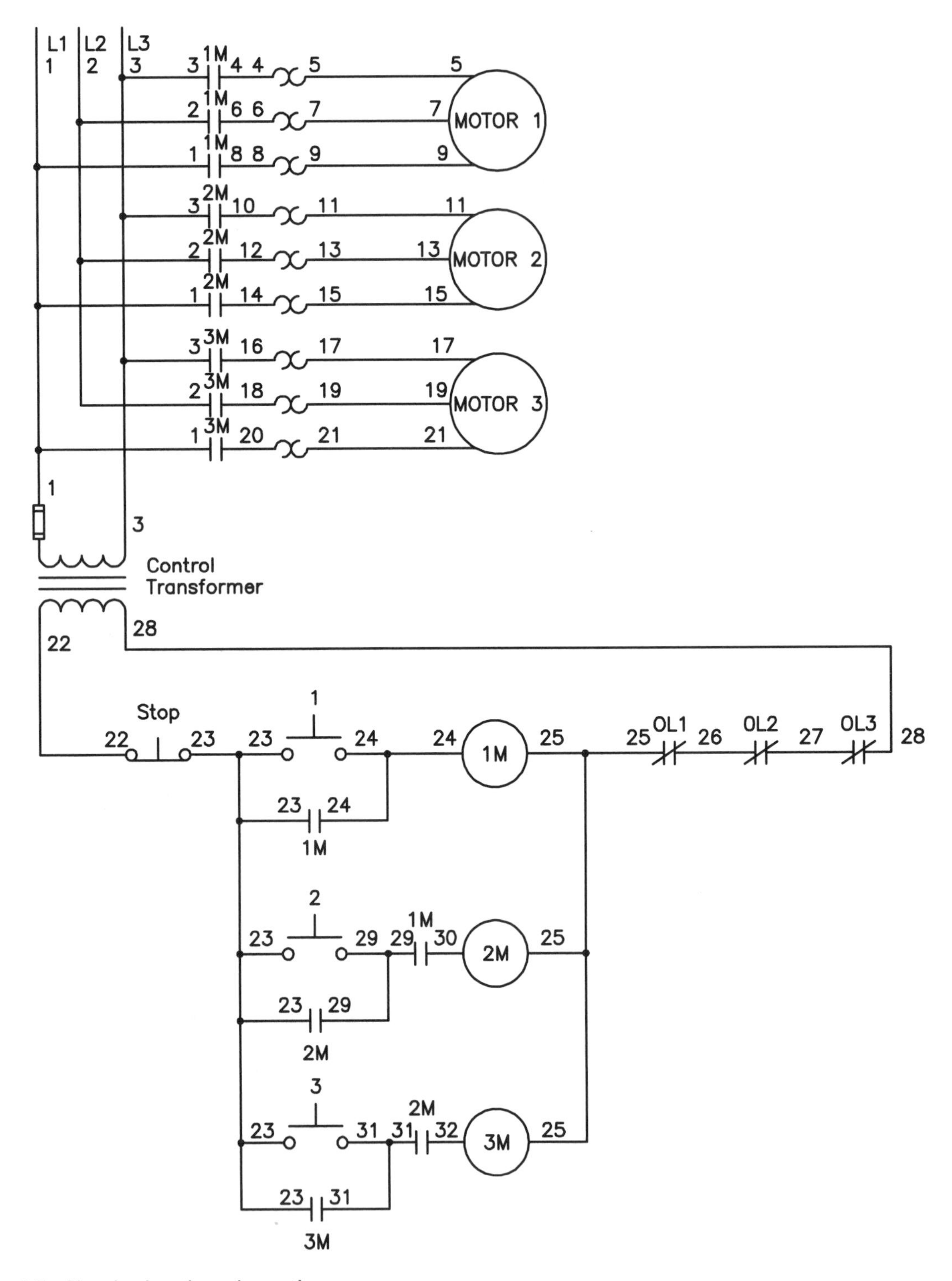

Figure 4-5 Numbering the schematic

4. Referring to Figure 4-5, assume that the normally open 2M contact located between wire numbers 23 and 29 is welded closed. Also assume that none of the motors is running. What would happen if
   a. The number 2 push button were to be pressed before the number 1 push button?

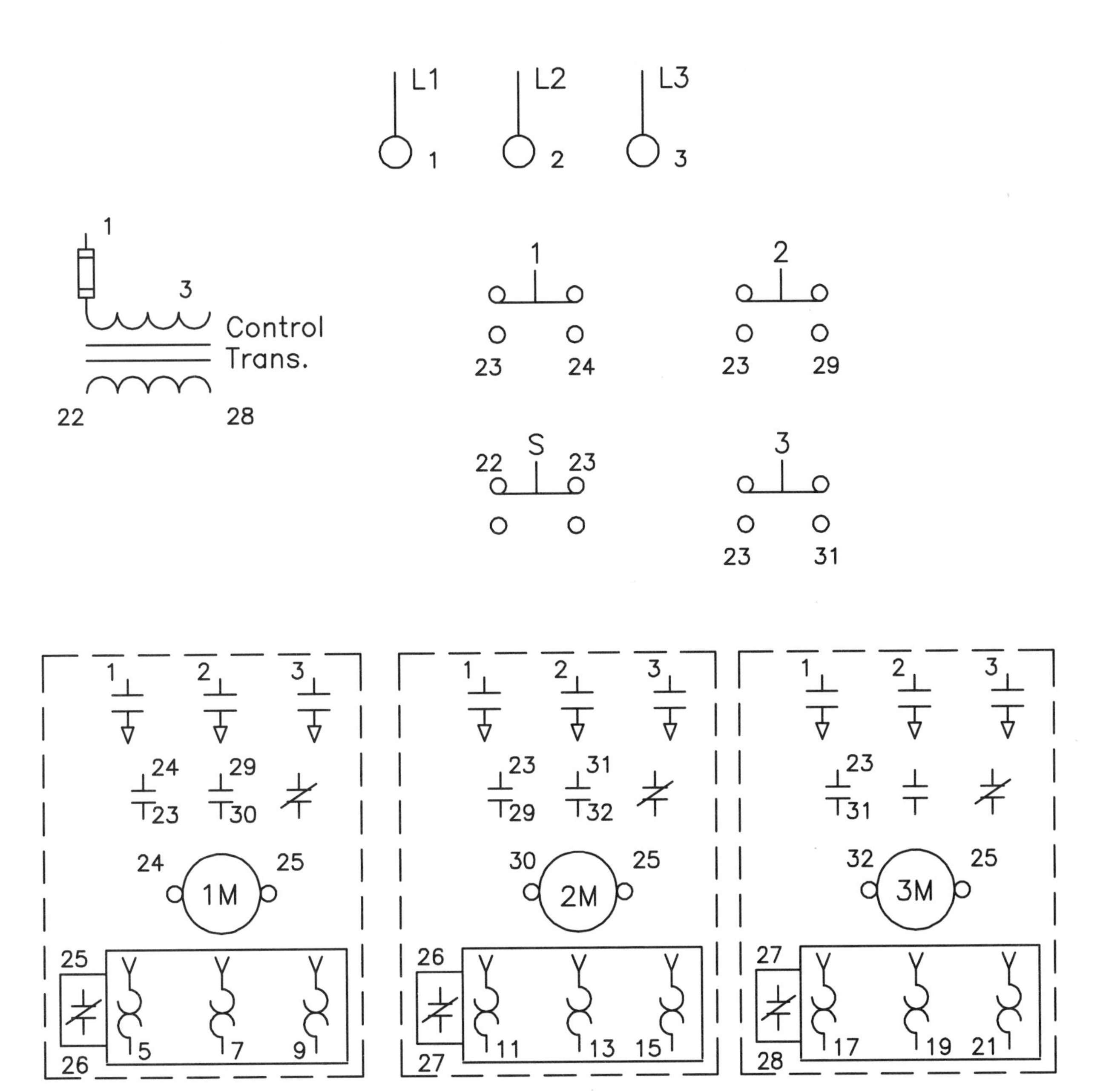

Figure 4-6 Numbering the components

b. The number 1 push button were to be pressed first?

Figure 4-7 Circuit redesign

5. In the control circuit shown in Figure 4-2, if an overload occurs on any motor all three motors will stop running. In the space provided in Figure 4-7, redesign the circuit so that the motors must still start in sequence from 1 to 3, but an overload on any motor will stop only that motor. If an overload should occur on motor 1, for example, motors 2 and 3 would continue to operate.

EXPERIMENT 5

Jogging Controls

Name:	Date:	Grade:
Comments:		

Objectives

After completing this experiment, the student should be able to:

- Describe the difference between inching and jogging circuits
- Discuss different jogging control circuits
- Draw a schematic diagram of a jogging circuit
- Discuss the connection of an 8 pin control relay
- Connect a jogging circuit in the laboratory using double acting push buttons
- Connect a jogging circuit in the laboratory using an 8 pin control relay

Materials Needed

- Three-phase power supply
- Three-phase motor starter
- One three-phase motor or equivalent motor load
- Three double acting push buttons (N.O./N.C. on each button)
- One 8 pin tube socket
- One 8 pin control relay
- One single-pole switch
- Control transformer

Jogging or inching control is used to help position objects by permitting the motor to be momentarily connected to power. Jogging and inching are very similar, and the terms are often used synonymously. Both involve starting a motor with short jabs of power. The difference between jogging and inching is that when a motor is jogged, it is started with short jabs of power at full voltage. When a motor is inched, it is started with short jabs at reduced power. Inching circuits requires the use of two contactors: one to run the motor at full power and the other to start the motor at reduced power, Figure 5-1. The run contactor is generally a motor starter that contains an overload relay, whereas the inching contactor does not. In the circuit shown in Figure 5-1, if the inch push button is pressed, a circuit is completed to S contactor coil causing all S contacts to close. This connects the motor to the line through a set of series resistors used to reduce power to the motor. Note that there is no S holding contact in parallel with the inch push button. When the push button is released, S contactor de-energizes and all S contacts reopen and disconnect the motor from the power line. If the run push button is pressed, M contactor energizes and connects the motor directly to the power line. Note the normally open M auxiliary contact connected in parallel with the run push button to maintain the circuit when the button is released.

Other Jogging Circuits

Like most control circuits, jog circuits can be connected in different ways. One method is shown in Figure 5-2. In this circuit a simple single-pole switch is inserted in series with the

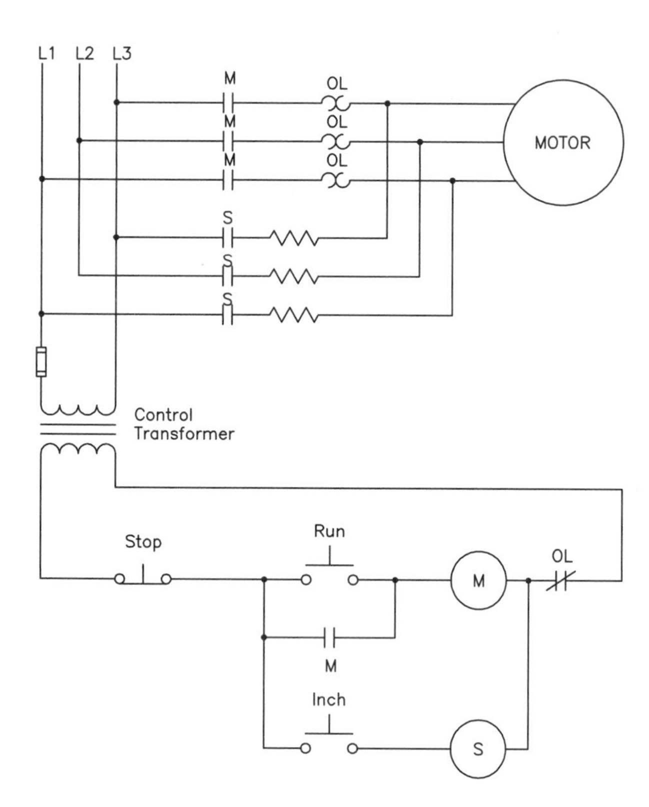

Figure 5-1 Inching control circuit

normally open M auxiliary contact connected in parallel with the start button. When the switch is open, it is in the **jog** position and prevents M holding contact from providing a complete path to M coil. When the start button is pushed, M coil will energize and connect the motor to the power line. When the start button is released, M coil will de-energize and disconnect the motor from the line. If the switch is closed, it is in the **run** position and permits the holding contact to complete a circuit around the start button.

Another method of constructing a run-jog control is shown in Figure 5-3. This circuit employs a double acting push button as the jog button. The normally closed section of the jog push button is connected in series with the normally open M auxiliary holding contact. If the jog button is pressed, the normally closed section of the button opens to disconnect the holding contacts before the normally open section of the button closes. Although M auxiliary contact closes when M coil energizes, the now open jog button prevents it from completing a circuit to the coil. When the jog button is released, the normally open section reopens and breaks contact before the normally closed section can reclose.

Although a double acting push button can be used to construct a run-jog circuit, it is not generally done because there is a possibility that the normally closed section of the jog button could reclose before the normally open section reopens. This could cause the holding contacts to lock the circuit in the run position, causing an accident. To prevent this possibility, a control relay is often employed, Figure 5-4. In the circuit shown in Figure 5-4, if the jog push button is pressed, M contactor energizes and connects the motor to the line. When the jog button is released, M coil de-energizes and disconnects the motor from the line.

When the run push-button is pressed, CR relay energizes and closes both CR contacts. The CR contacts connected in parallel with the run button close to maintain the circuit to CR coil, and the CR contacts connected in parallel with the jog button close and complete a circuit to M coil.

Connecting Jogging Circuits

In this experiment, four different jog circuits will be connected in the laboratory. Three of these circuits are illustrated in Figures 5-2, 5-3, and 5-4. The fourth circuit will be designed by the student in accord with given circuit parameters.

Connecting Circuit 1

1. Refer to the schematic diagram in Figure 5-2. Place wire numbers beside the components following the procedure discussed in previous experiments.
2. Using the components shown in Figure 5-5, place corresponding wire numbers beside the components.
3. Connect the circuit by following the wire numbers in the schematic diagram in Figure 5-2.
4. After checking with the instructor, turn on the power and test the circuit for proper operation. The motor should jog when the switch is open and run when the switch is closed.
5. Turn off the power and disconnect the circuit.

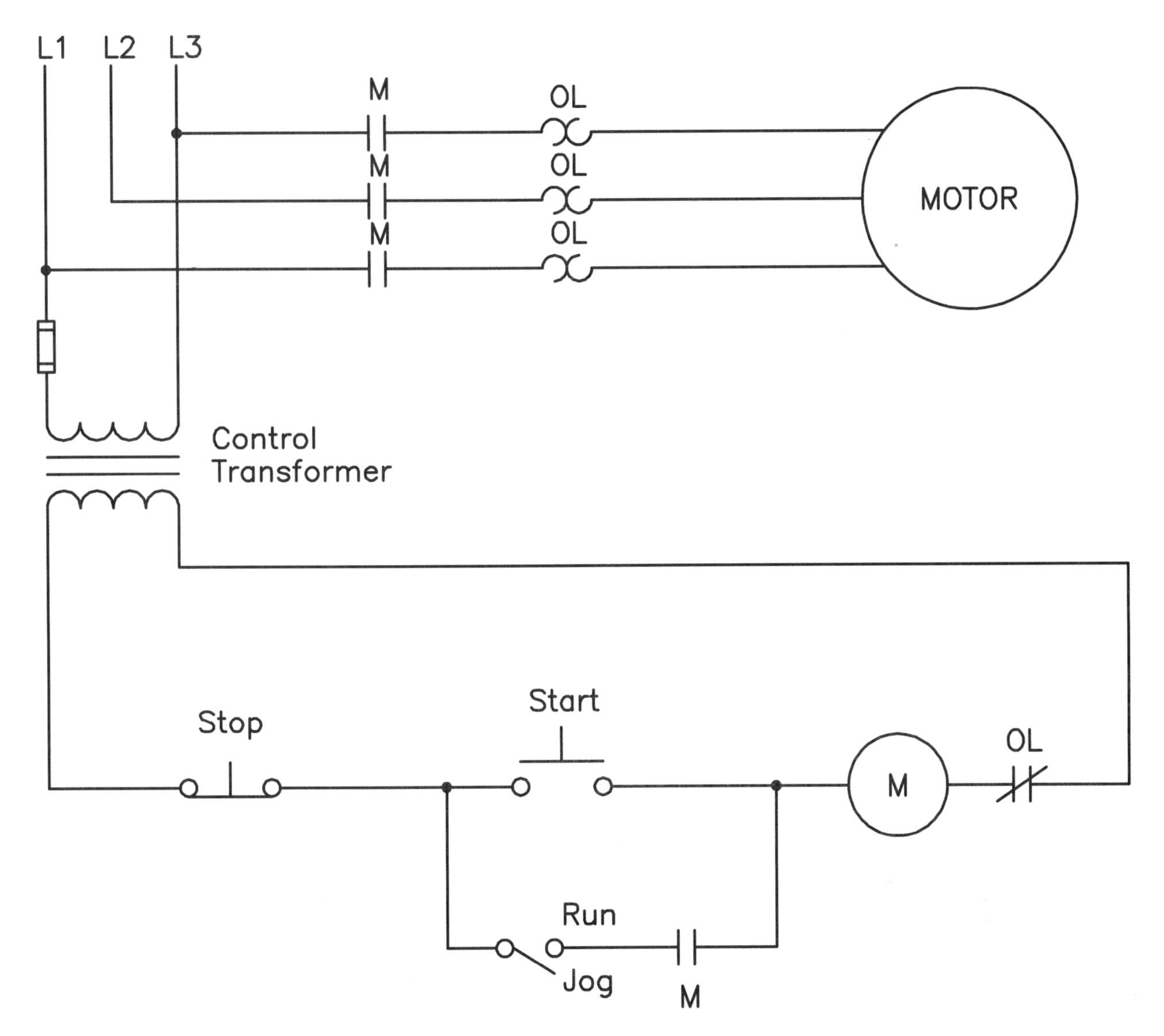

Figure 5-2 Run-jog control using a single pole switch

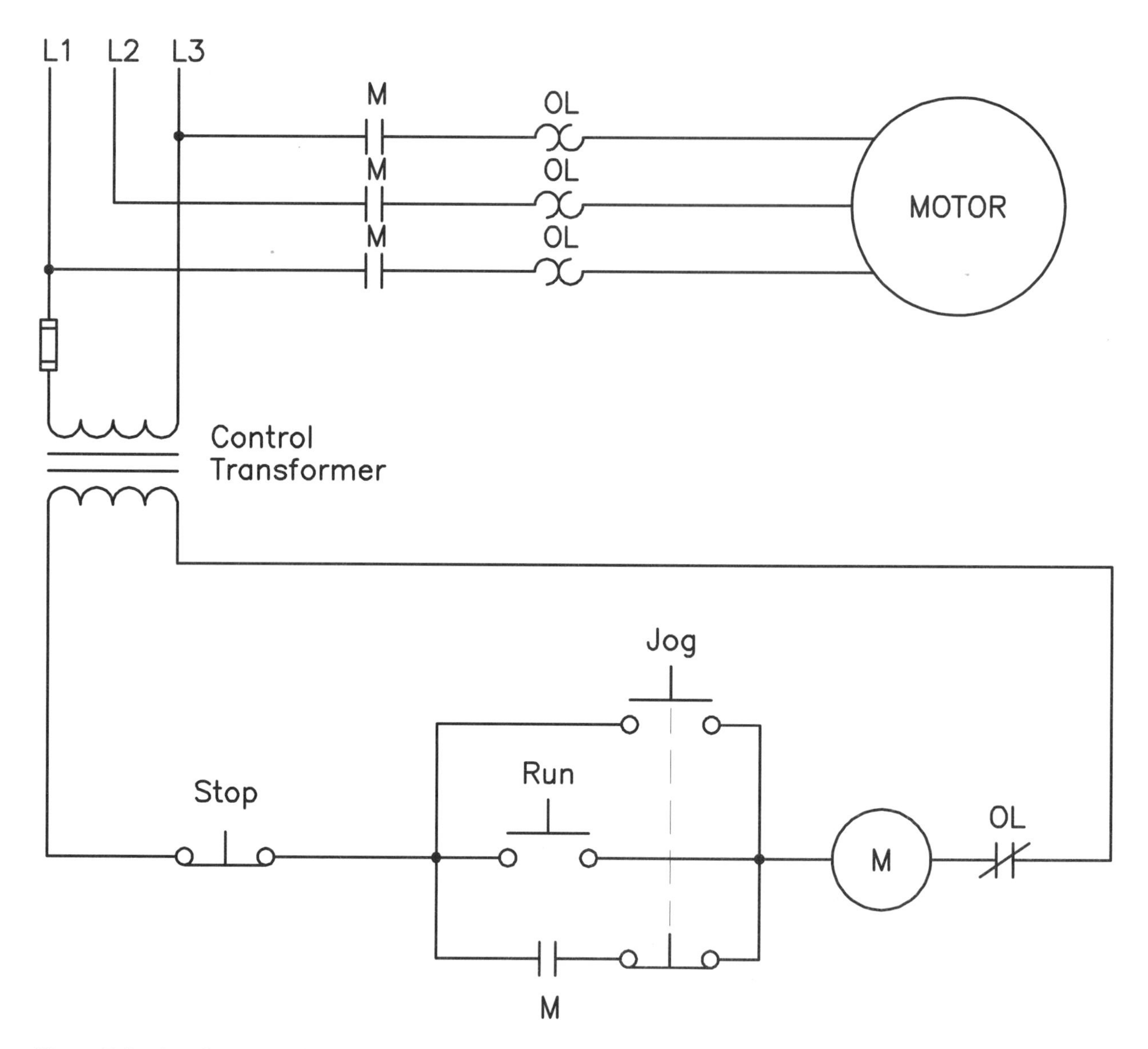

Figure 5-3 Jogging control using a double acting push button

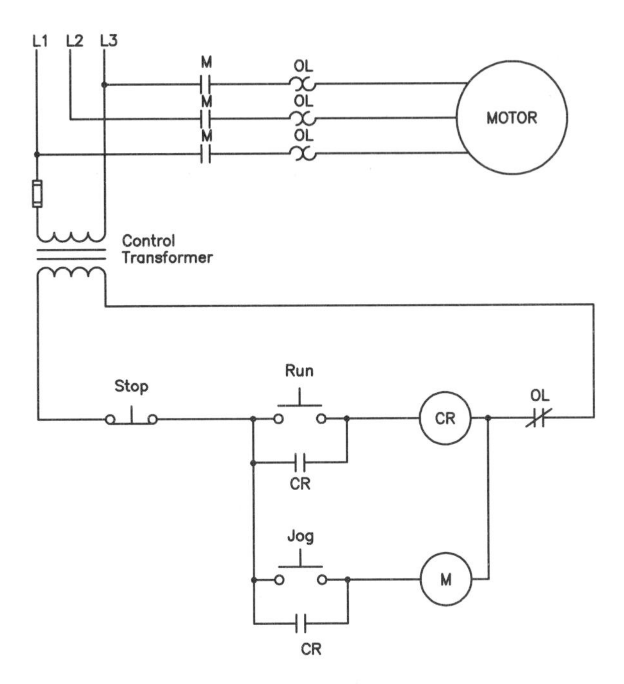

Figure 5-4 Run-jog control using a control relay

Connecting the Second Run-Jog Circuit

1. Using the schematic shown in Figure 5-3, place wire numbers beside the components.
2. Place corresponding wire numbers beside the components shown in Figure 5-6.
3. Connect the circuit using the schematic diagram in Figure 5-3.
4. After checking with the instructor, turn on the power and test the circuit for proper operation.
5. Turn off the power and disconnect the circuit.

Connecting the Third Run-Jog Circuit

The third run-jog circuit involves the use of a control relay. In this circuit, an 8 pin control relay will be used. Eight pin relays are designed to fit into an 8 pin tube socket. Therefore the socket is the device to which connection is made, not the relay itself. Eight pin relays commonly have coils with different voltage ratings such as 12 VDC, 24 VDC, 24 VAC, and 120 VAC, so make certain that the coil of the relay you use is rated for the circuit control voltage. Most 8 pin relays contain two single-pole double-throw contacts. A diagram showing the standard pin connection for 8 pin relays with two sets of contacts is shown in Figure 5-7.

Connecting the Tube Socket

When making connections to tube sockets, it is generally helpful to place the proper relay pin numbers beside the component on the schematic diagram. To distinguish pin numbers from wire numbers, pin numbers will be circled. The schematic in Figure 5-4 is shown in Figure 5-8 with the addition of relay pin numbers. The connection diagram in Figure 5-7 shows that the relay coil is connected to pins 2 and 7. Note that CR relay coil in Figure 5-8 has a circled 2 and 7 placed beside it.

The connection diagram also indicates that the relay contains two sets of normally open contacts. One set is connected to pins 1 and 3, and the other set is connected to pins 8 and 6. Note in the schematic of Figure 5-8 that one of the normally open CR contacts has the circled numbers 1 and 3 beside it and the other normally open CR contact has the circled numbers 8 and 6 beside it.

1. Using the drawing in Figure 5-8, place wire numbers on the schematic.
2. Using the wire numbers placed on the schematic diagram in Figure 5-8, place corresponding wire numbers beside the proper components shown in Figure 5-9.
3. Connect the circuit shown in Figure 5-8.
4. After checking with the instructor, turn on the power and test the circuit for proper operation.
5. Turn off the power and disconnect the circuit.

Review Questions

1. Explain the difference between inching and jogging.

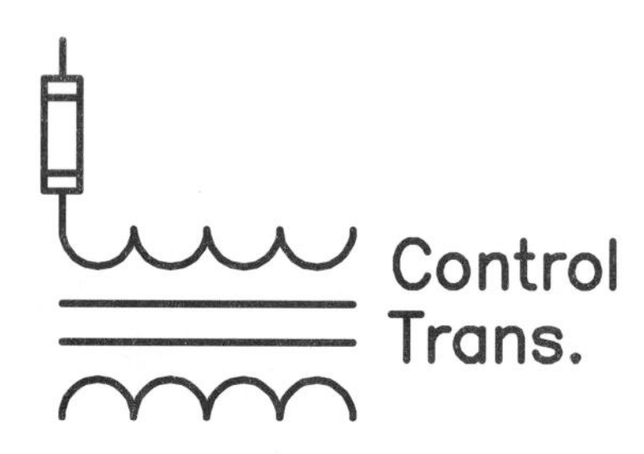

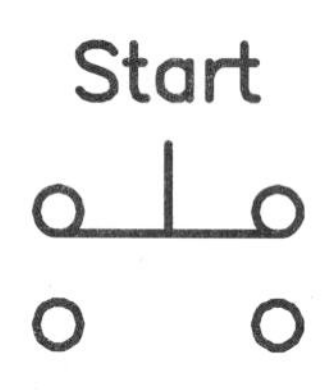

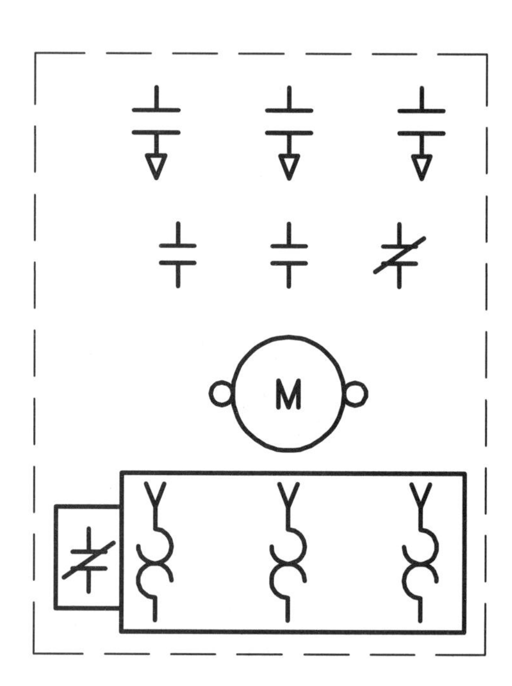

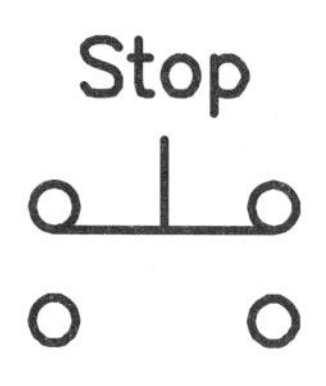

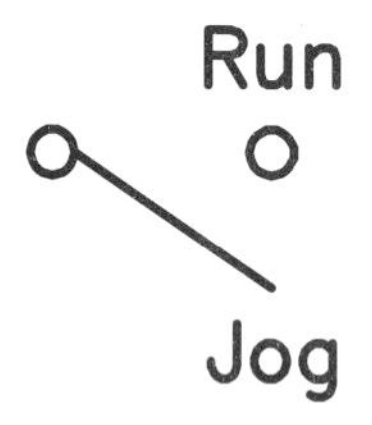

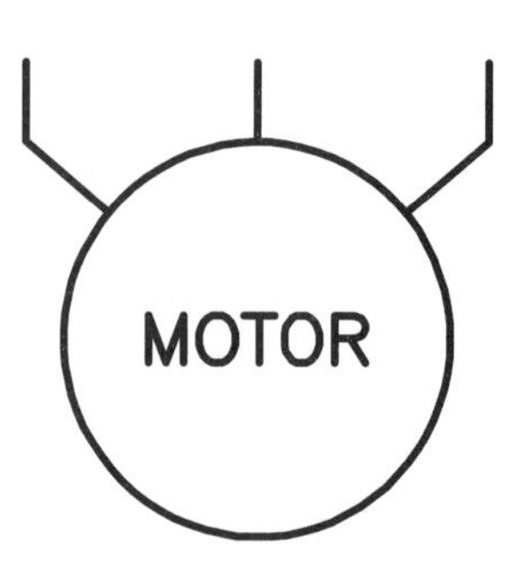

Figure 5-5 Components needed to connect circuit 1

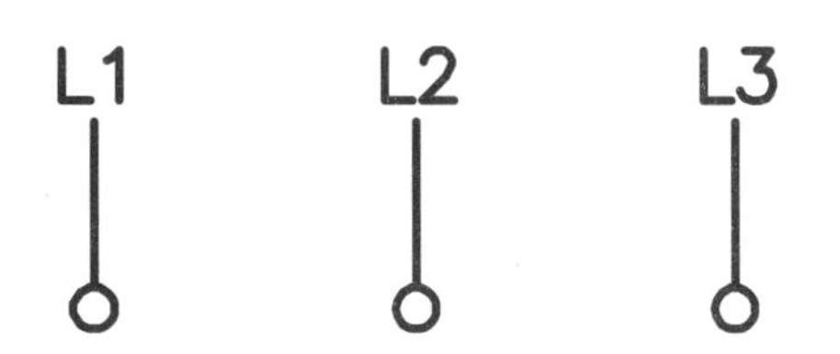

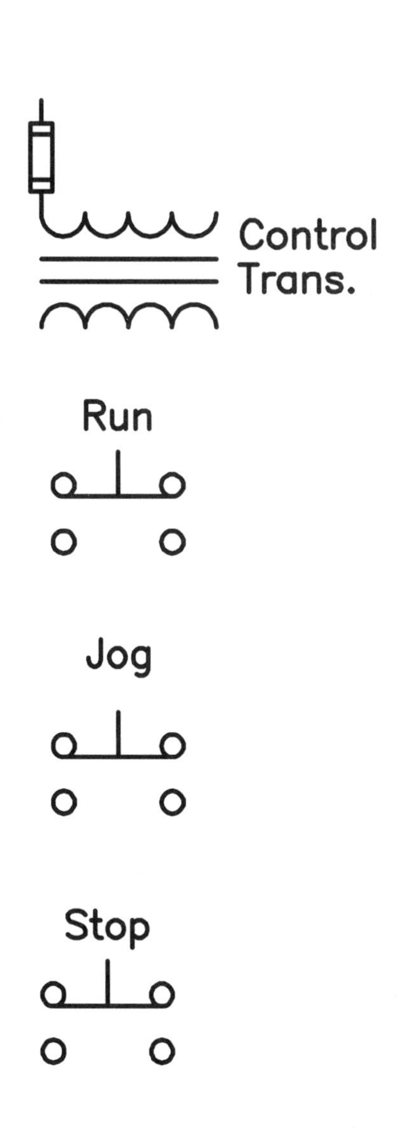

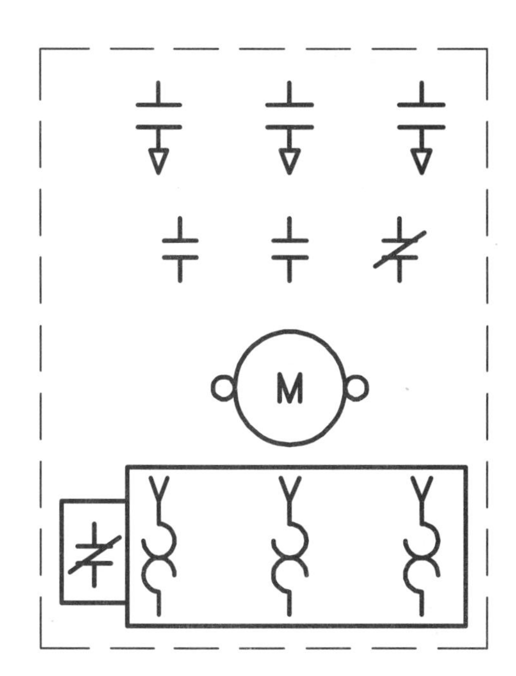

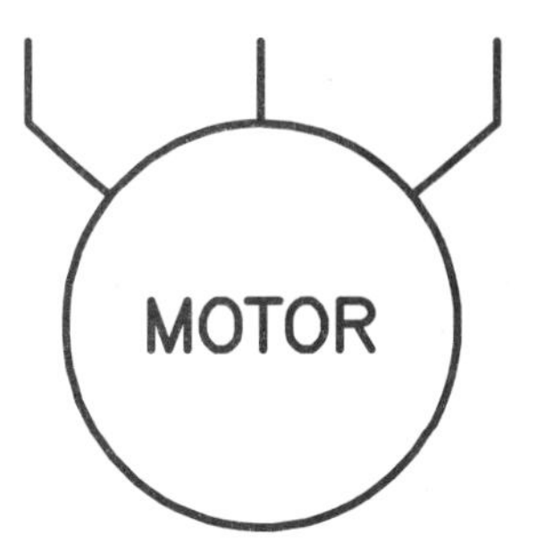

Figure 5-6 Components needed to connect the second run-jog circuit

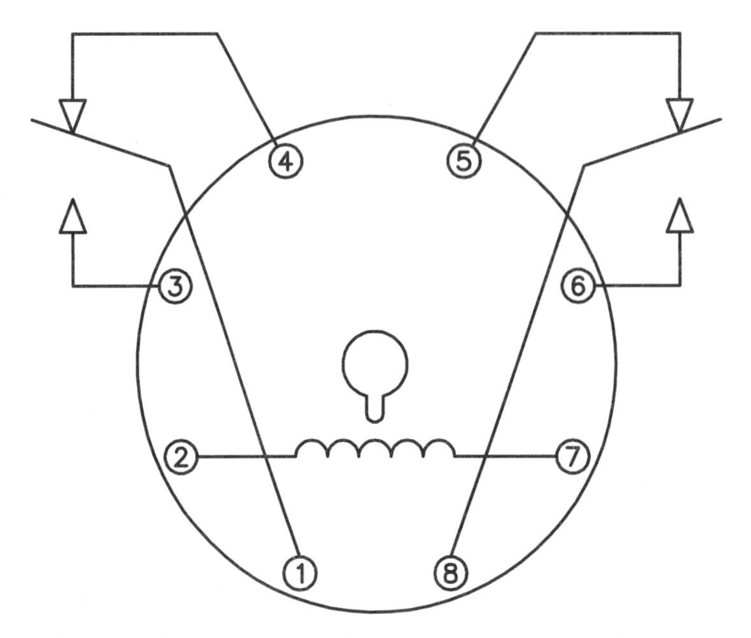

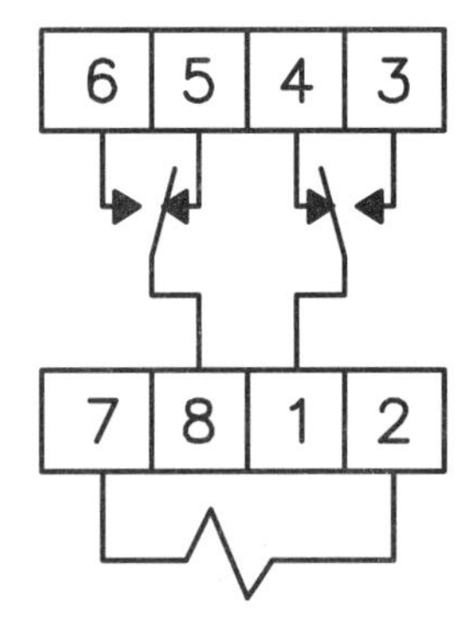

Figure 5-7 Standard diagram for an 8 pin control relay

2. What is the main purpose of jogging?

3. Refer to the circuit shown in Figure 5-10. In this circuit, the jog button has been connected incorrectly. The normally closed section has been connected in parallel with the run push button, and the normally open section has been connected in series with the holding contacts. Explain how this circuit operates.

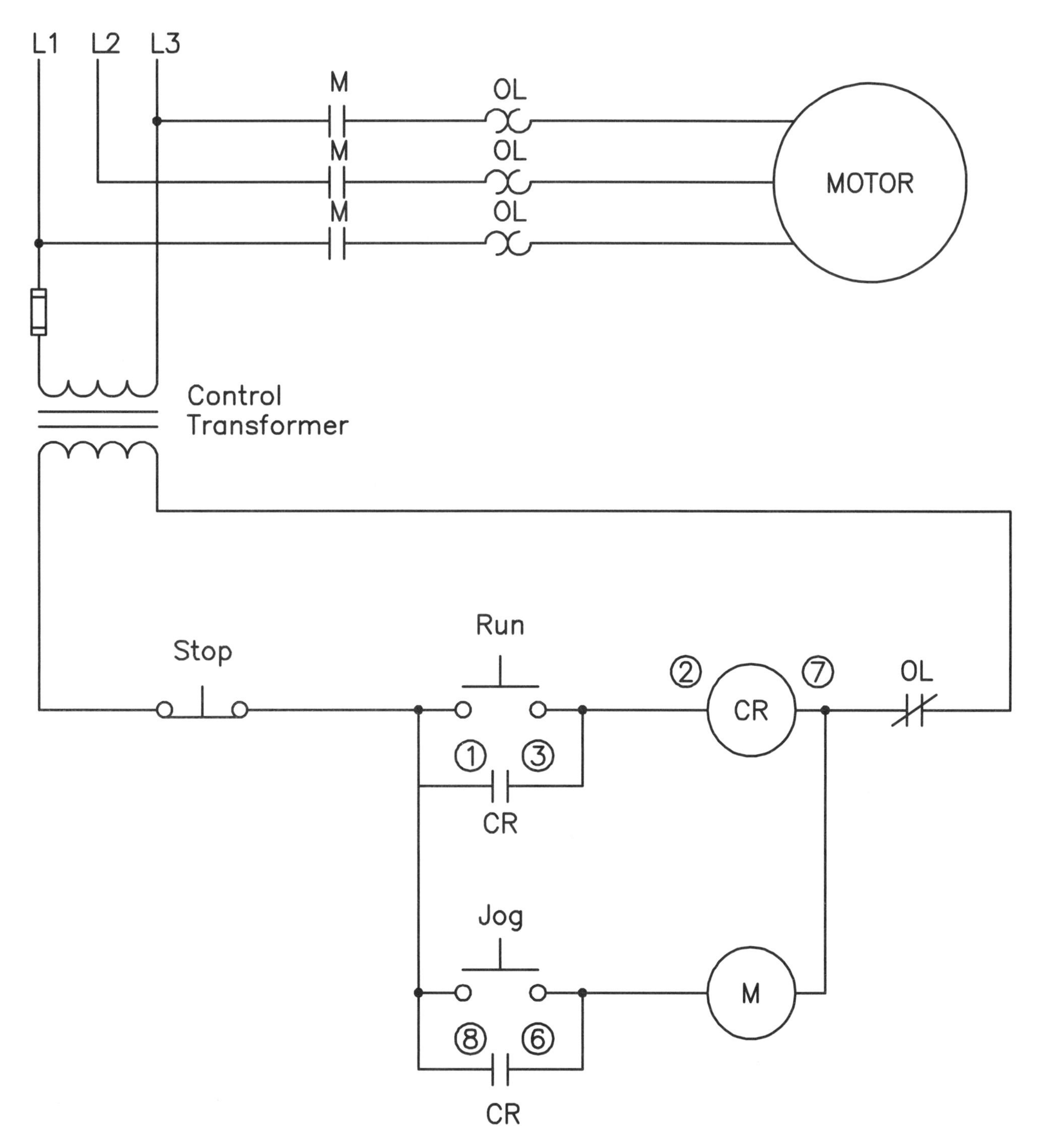

Figure 5-8 Adding pin numbers aids in connecting the circuit

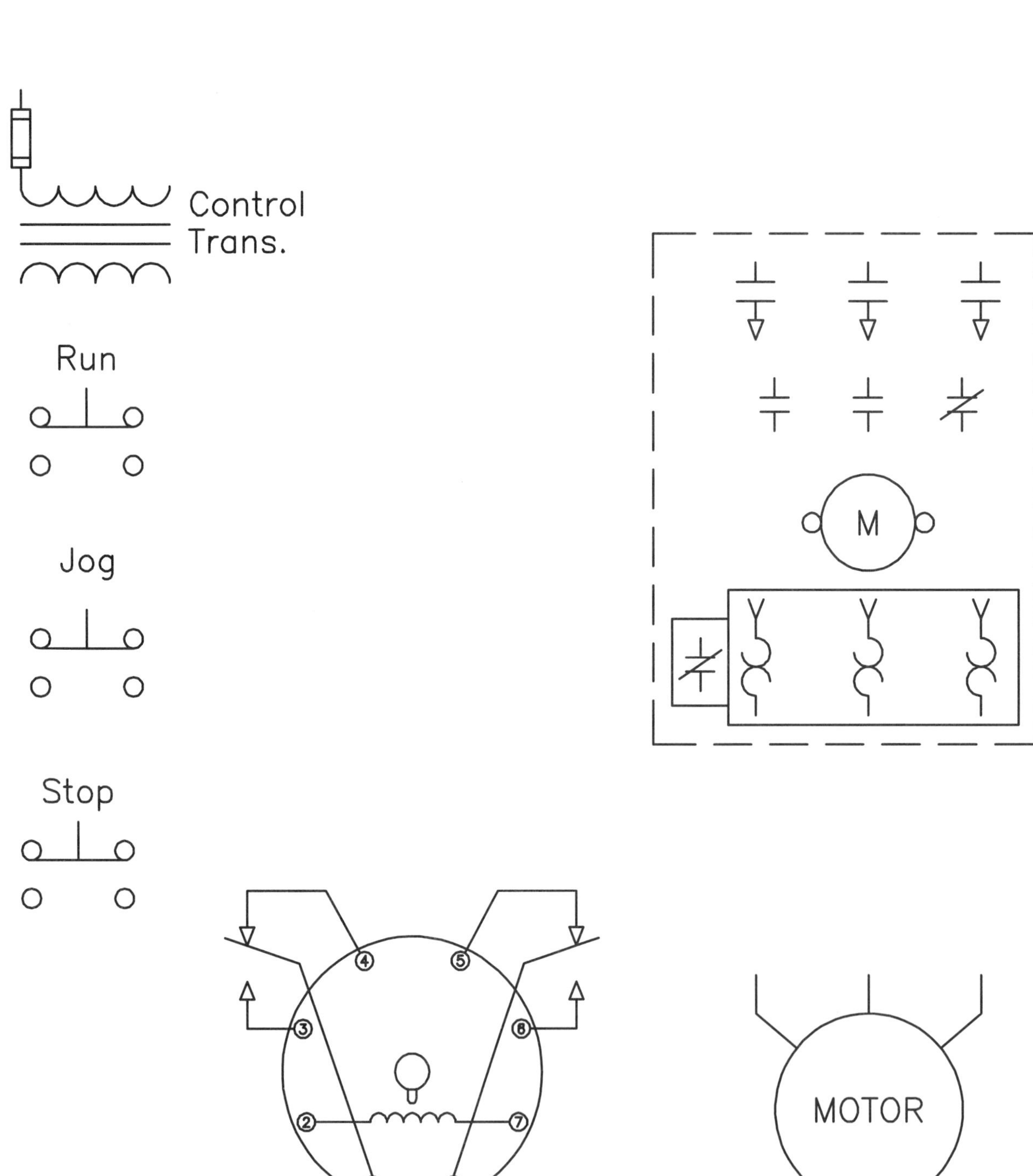

Figure 5-9 Components needed to connect circuit 3

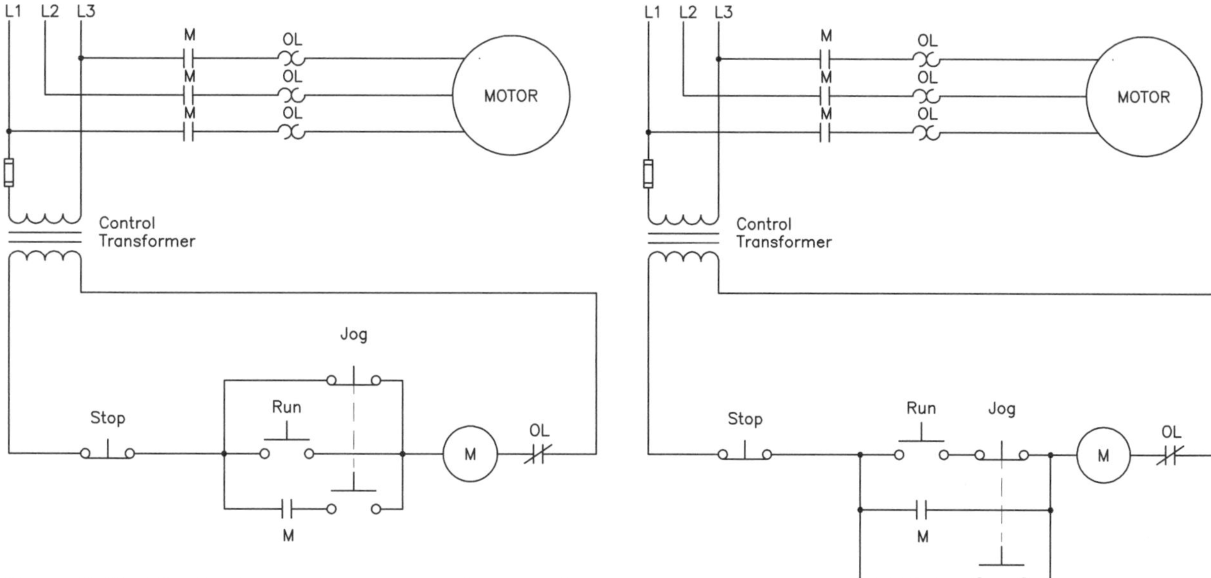

Figure 5-10 The jog button is connected incorrectly

Figure 5-11 Another incorrect connection for the jog button

4. Refer to the circuit shown in Figure 5-11. In this circuit the jog push button has again been connected incorrectly. The normally closed section of the button has been connected in series with the normally open run push button, and the normally open section of the jog button is connected in parallel with the holding contacts. Explain how this circuit operates.

5. In the space provided in Figure 5-12, design a run-jog circuit to the following specifications.
   a. The circuit contains two push buttons: a normally closed stop button and a normally open start button.
   b. When the start button is pressed, the motor will run normally. When the stop button is pressed, the motor will stop.
   c. If the stop button is manually held in, however, the motor can be jogged by pressing the start button.
   d. The circuit contains a control transformer, motor, and three-phase motor starter with at least one normally open auxiliary contact.

6. After your instructor has approved the new circuit design, connect the circuit in the laboratory.

7. Turn on the power and test the circuit for proper operation.

8. Turn off the power and disconnect the circuit. Return the components to their proper place.

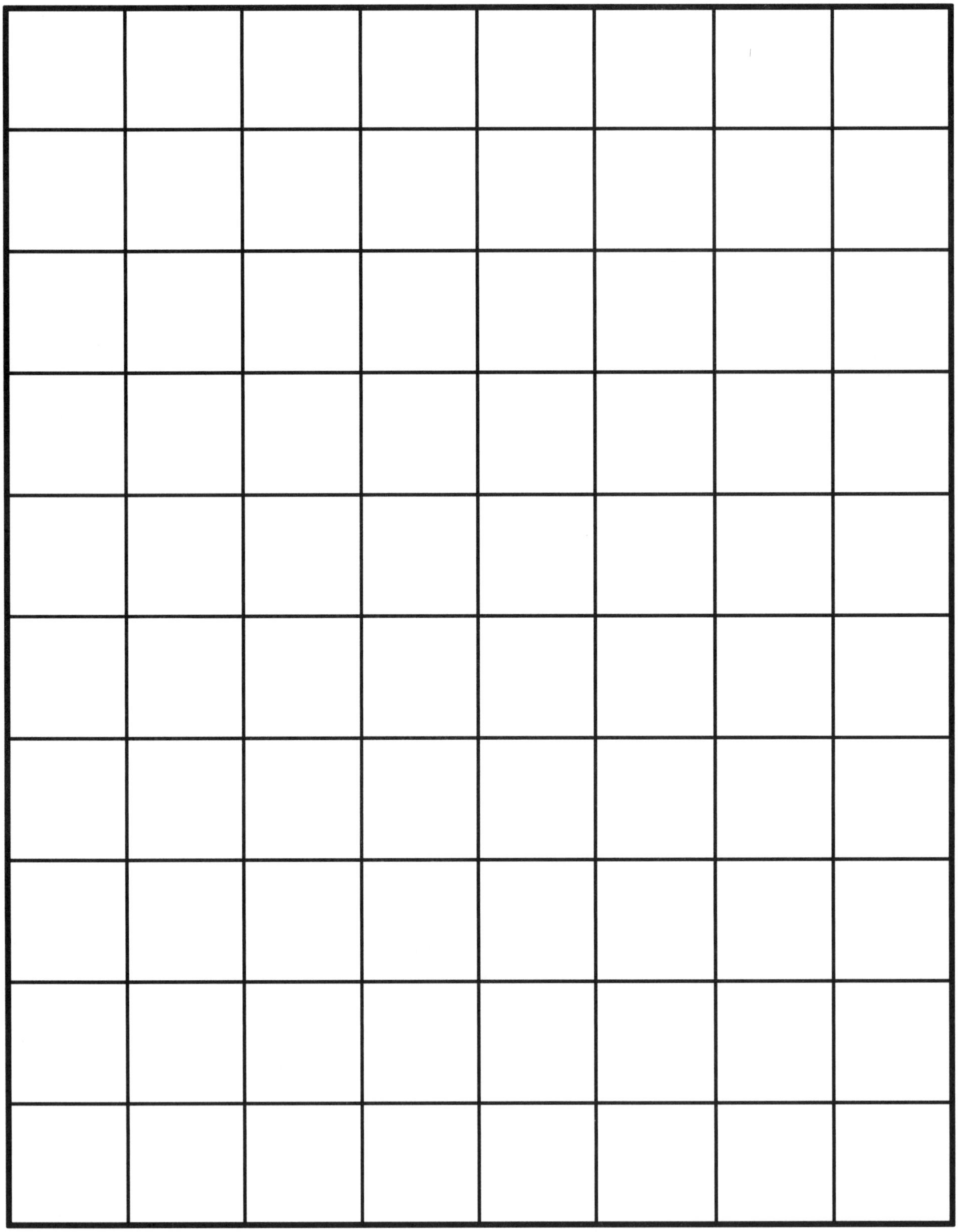

Figure 5-12 Circuit design

EXPERIMENT 6

On-Delay Timers

Name:	Date:	Grade:
Comments:		

Objectives

After completing this experiment, the student should be able to:

- Discuss the operation of an on-delay timer
- Draw the NEMA contact symbols used to represent both normally open and normally closed on-delay contacts
- Discuss the difference in operation between pneumatic and electronic timers
- Connect a circuit in the laboratory employing an on-delay timer

Materials Needed

- Three-phase power supply
- Control transformer
- Two double acting push buttons (N.O./N.C. on each button)
- 2 ea.—Three-phase motor starter with at least one normally open auxiliary contact
- Dayton Solid-State Timer—model 6A855 or equivalent and 11 pin socket
- 8 pin control relay and 8 pin socket
- 2 ea.—Three-phase motors or equivalent motor loads

Timers can be divided into two basic types: on-delay and off-delay. Although there are other types such as one shot and interval, they are basically an on- or off-delay timer. In this unit, the operation of on-delay timers will be discussed. The operating sequence of an on-delay timer is as follows.

When the coil is energized, the timed contacts will delay changing position for some period of time. When the coil is de-energized, the timed contacts will return to their normal position immediately. In this explanation, the word **timed** contacts is used. The reason is that some timers contain both timed and instantaneous contacts. When using a timer of this type, care must be taken to connect to the proper set of contacts.

Timed Contacts

The timed contacts are controlled by the action of the timer, whereas the instantaneous contacts operate like any standard set of contacts on a control relay; when the coil energizes, the contacts change position immediately, and when the coil de-energizes, they change back to their normal position immediately.

The standard NEMA symbols used to represent on-delay contacts are shown in Figure 6-1. The arrow points in the direction the contact will move after the delay period. The normally open contact, for example, will close after the time delay period, and the normally closed contact will open after the time delay period.

Instantaneous Contacts

Instantaneous contacts are drawn in the same manner as standard relay contacts. Figure 6-2 illustrates a set of instantaneous contacts controlled by timer TR. The instantaneous contacts are often used as holding or sealing contacts in a control circuit. The control circuit

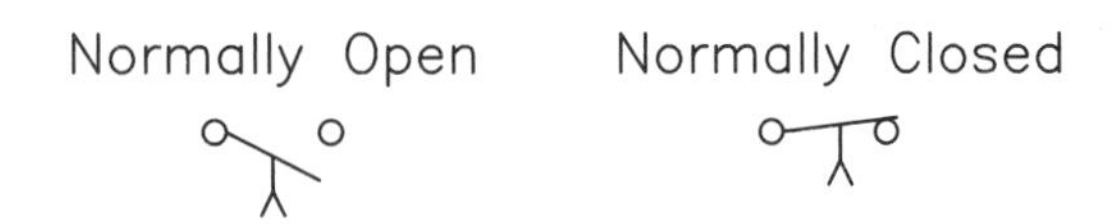

Figure 6-1 NEMA standard symbols for on-delay contacts

Normally Open Normally Closed

TR TR

Figure 6-2 Instantaneous contact symbols

shown in Figure 6-3 illustrates an on-delay timer used to delay the starting of a motor. When the start push button is pressed, TR coil energizes and the normally open instantaneous TR contacts close immediately to hold the circuit. After the preset time period, the normally open TR timed contacts will close and energize the coil of M starter, which connects the motor to the line.

When the stop button is pressed and TR coil de-energizes, both TR contacts return to their normal position immediately. This de-energizes M coil and disconnects the motor from the line.

Control Relays Used with Timers

Not all timers contain instantaneous contacts. Most electronic timers, for example, do not. When an instantaneous contact is needed and the timer does not have one available, it is common practice to connect the coil of a control relay in parallel with the coil of the timer, Figure 6-4. In this way the electronic timer will operate with the timer. In the circuit shown in Figure 6-4, both coils TR and CR will energize when the start button is pressed. This causes CR contact to close and seal the circuit.

Time Delay Methods

Although there are two basic types of timers, there are different methods employed to obtain a time delay. One of the oldest methods that is still in general use is the pneumatic timer. Pneumatic timers use a bellows or diaphragm and operate on the principle of air displacement. Some type of needle valve is generally used to regulate the airflow and thereby regulate the time delay. Pneumatic timers are simple in that they contain a coil, contacts, and some method of adjusting the amount of time delay. Because of their simplicity

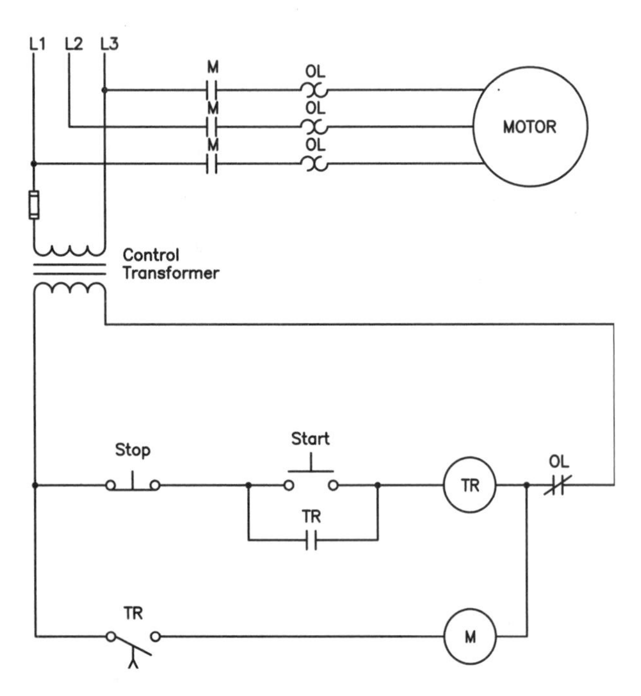

Figure 6-3 The motor starts after the start button is pressed

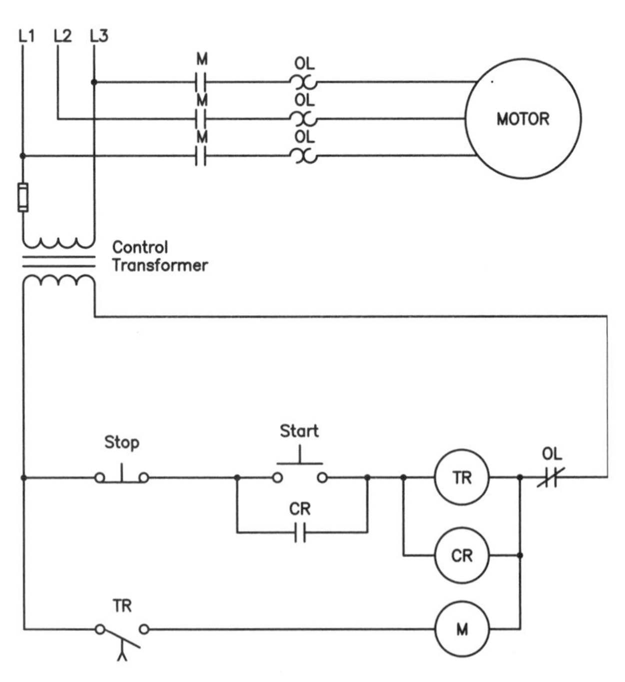

Figure 6-4 A control relay furnishes the instantaneous contact

of operation, when control circuits are in the design stage, the circuit logic is generally developed with the assumption that pneumatic timers will be used. After the circuit logic has been developed, it may be necessary to make changes that will accommodate a particular type of timer.

Another very common method of providing a time delay is with an electric clock similar to a wall clock. These timers contain a small single-phase synchronous motor. As a general rule, most clock timers can be set for different full-scale values by changing the gear ratio.

Electronic timers are becoming very popular for several reasons:

1. They are much less expensive than pneumatic or clock timers.
2. They have better repeat accuracy than pneumatic or clock timers.
3. Most can be set for 0.1 second delays, and many can be set to an accuracy of 0.01 second.
4. Many electronic timers are intended to be plugged into an 8 or 11 pin tube socket. This makes replacing the timer much simpler and takes less time.

The First Circuit

The first circuit to be connected is shown in Figure 6-4. In this circuit it will be assumed that an 11 pin timer is being used and that the coil is connected to pins 2 and 10, and a set of normally open timed contacts are connected to pins 1 and 3. The coil of the 8 pin control relay is connected to pins 2 and 7 and a normally open contact is connected to pins 1 and 3. When using control devices that are connected with 8 and 11 pin sockets, it is generally helpful to place pin numbers beside the component. To prevent pin numbers from being confused with wire numbers, a circle will be drawn around the pin numbers, Figure 6-5.

Connecting Circuit 1

1. Using the circuit shown in Figure 6-5, place wire numbers beside the components.
2. Connect the control part of the circuit by following the wire numbers placed beside the components. Note the pin numbers beside the coils and contacts of the timer and control relay.
3. Plug the timer and control relay into their appropriate sockets. Set the timer to operate as an on-delay timer and set the time period for 5 seconds.
4. After checking with the instructor, turn on the power and test the operation of the circuit.
5. Turn off the power.
6. If the control part of the circuit operated correctly, connect the motor or equivalent motor load.
7. Turn on the power and test the total circuit for proper operation.
8. Turn off the power and disconnect the circuit.

Discussing Circuit 2

In the next circuit, two motors are to be started with a five-second time delay between the starting of tvvhe first motor and the second motor. In this circuit a normally open auxiliary contact on starter 1M is used as the holding contact making the use of the control relay unnecessary.

When the start button is pressed, coils 1M and TR energize immediately. This causes motor 1 to start operating and timer TR to begin timing. After five seconds, TR contacts close and connect motor 2 to the line. When the stop button is pressed, or if an overload on either motor should occur, all coils will be de-energized and both motors will stop.

Connecting Circuit 2

1. Using the circuit shown in Figure 6-6, place pin numbers beside the timer coil and normally open contact.
2. Place wire numbers on the circuit in Figure 6-6.
3. Connect the control part of the circuit.
4. Turn on the power and test it for proper operation.
5. Turn off the power.
6. If the control part of the circuit operated properly, connect the motors or equivalent motor loads.
7. Turn on the power and test the circuit for proper operation.
8. Turn off the power and disconnect the circuit.

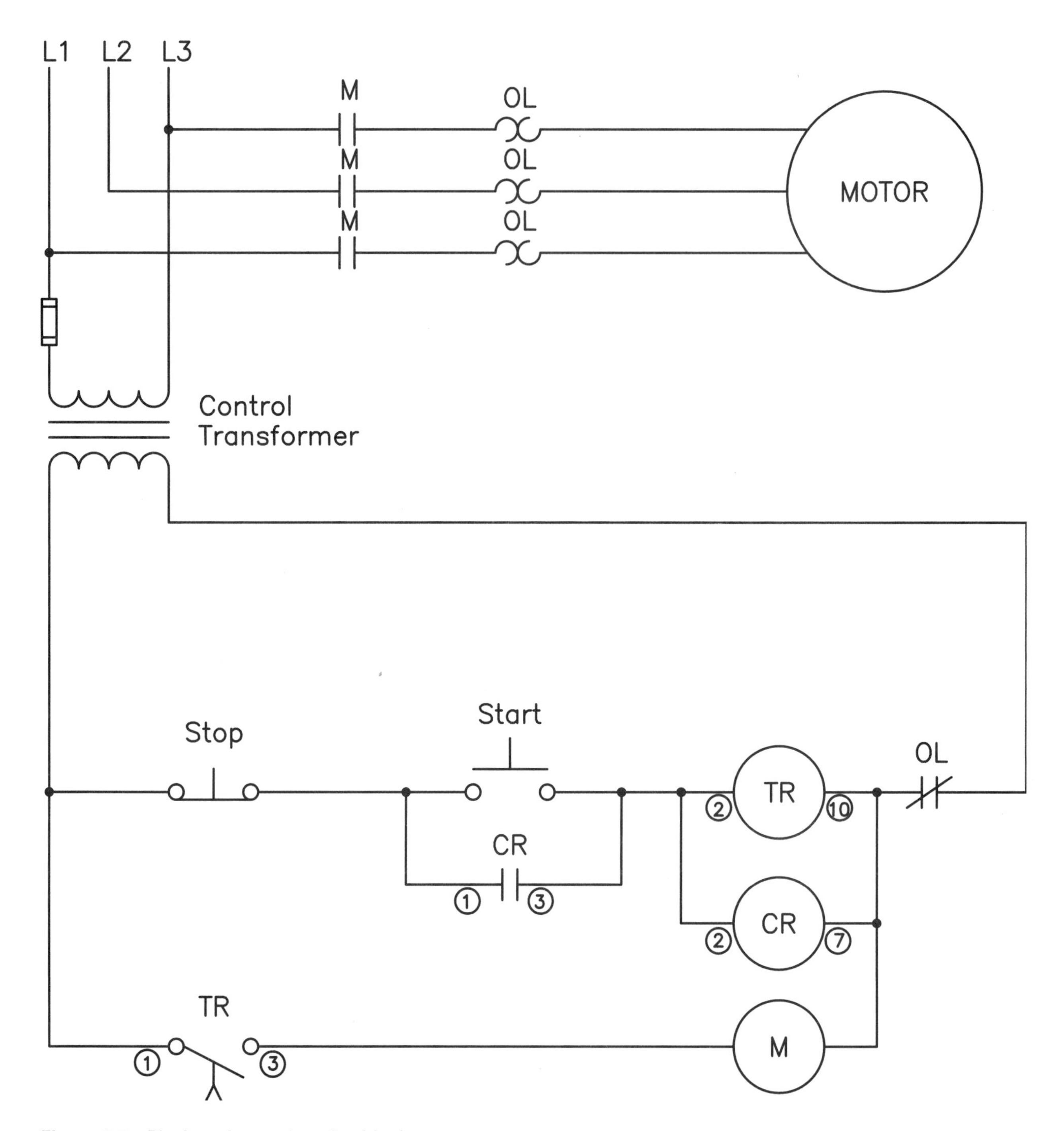

Figure 6-5 Placing pin numbers beside the components

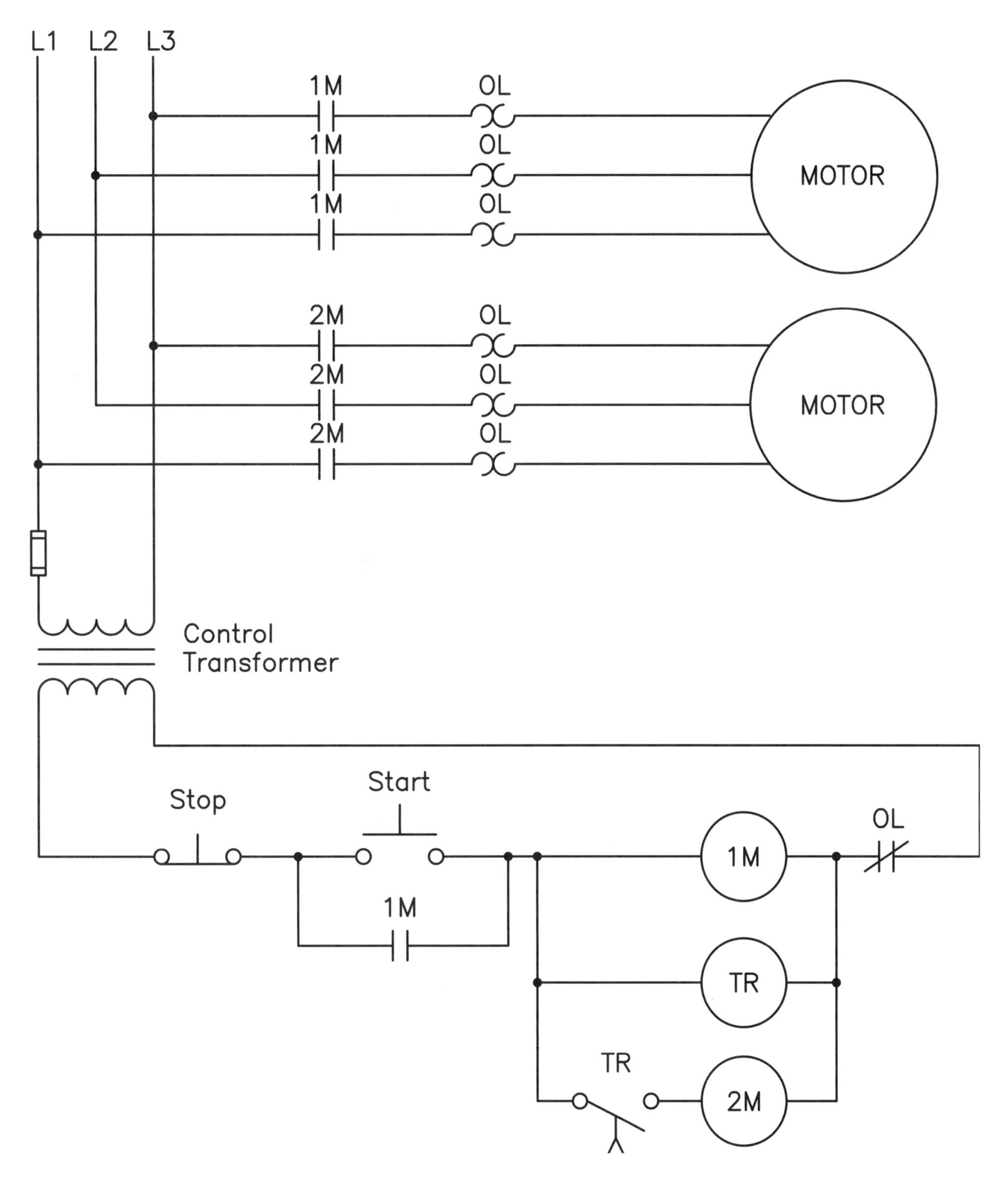

Figure 6-6 Motor 2 starts after motor 1

Review Questions

1. Explain the operation of an on-delay timer.

2. Explain the difference between timed contacts and instantaneous contacts.

3. Refer to the circuit shown in Figure 6-3. If the timer has been set for a delay of ten seconds, explain the operation of the circuit when the start button is pressed.

4. In the circuit shown in Figure 6-3, is it necessary to hold the start button closed for a period of at least ten seconds to ensure that the circuit will remain energized? Explain your answer.

5. Assume that the timer in Figure 6-3 is set for a delay of ten seconds. Now assume that the start button is pressed, and after a delay of eight seconds the stop button is pressed. Will the motor start two seconds after the stop button was pressed?

6. What is generally done to compensate when a set of instantaneous timer contacts is needed and the timer does not contain the set?

7. Refer to the circuit shown in Figure 6-6. Assume that it is necessary to stop the operation of both motors after the second motor has been operating for a period of ten seconds. Using the space provided in Figure 6-7, redraw the circuit to turn off both motors after the second motor has been in operation for ten seconds. (***Note:*** It will be necessary to use a second timer.)

8. After your instructor has approved the design change, connect the new circuit in the laboratory and test it for proper operation.

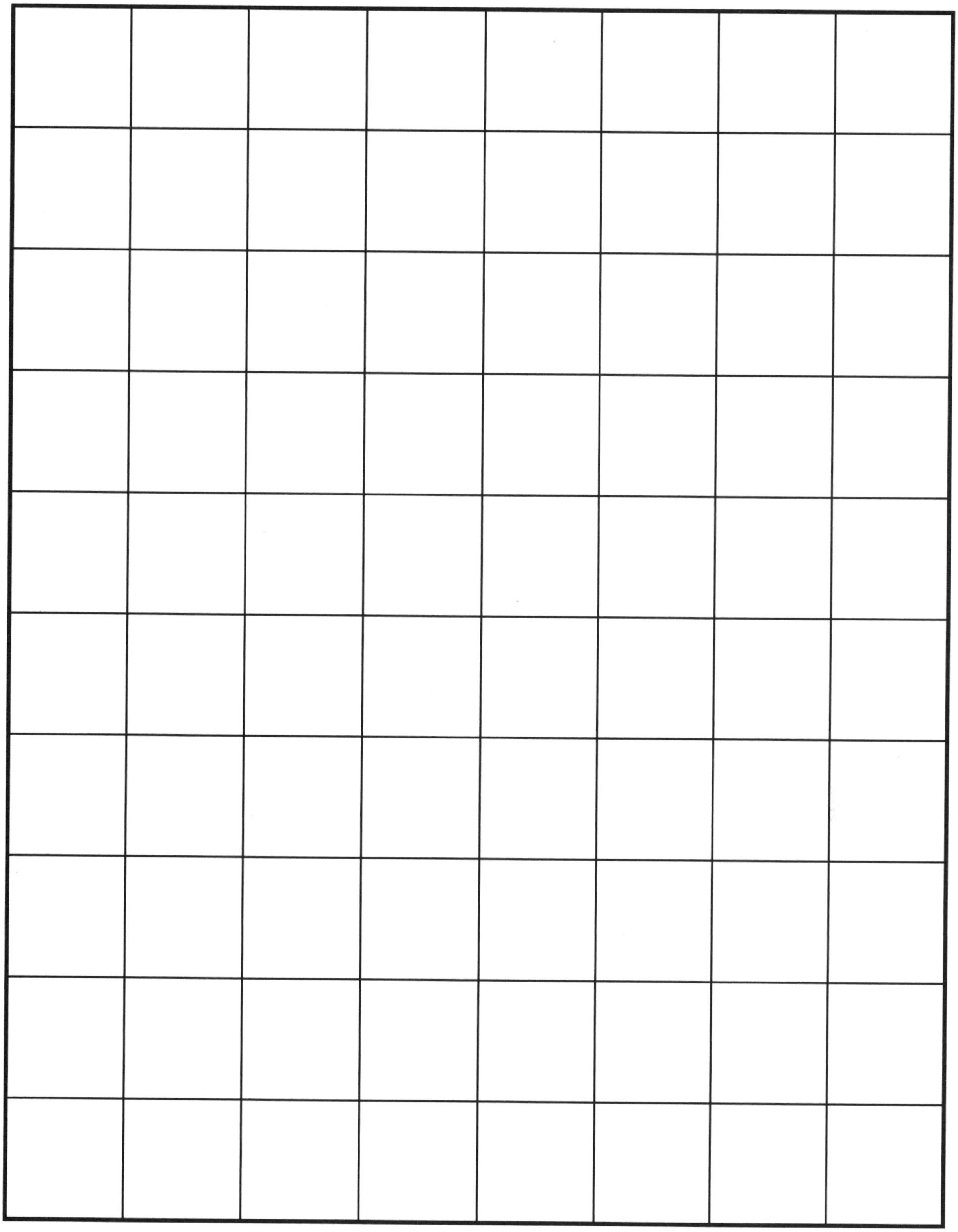

Figure 6-7 Circuit redesign

EXPERIMENT 7

Off-Delay Timers

Name: Date: Grade:

Comments:

Objectives

After completing this experiment, the student should be able to:

- Discuss the operation of an off-delay timer
- Draw the NEMA contact symbols used to represent both normally open and normally closed off-delay contacts
- Discuss the difference in operation between pneumatic and electronic timers
- Connect a circuit in the laboratory employing an off-delay timer

Materials Needed

- Three-phase power supply
- Control transformer
- Two double acting push buttons (N.O./N.C. on each button)
- 2 ea.—Three-phase motor starter with at least one normally open auxiliary contact
- Dayton Solid-State Timer—model 6A855 or equivalent
- 11 pin control relay and two 11 pin sockets
- 2 ea.—Three-phase motors or equivalent motor loads

The logic of an **off-delay** timer is as follows: **When the coil is energized, the timed contacts change position immediately. When the coil is de-energized the timed contacts remain in their energized position for some period of time before changing back to their normal position.** Figure 7-1 shows the standard NEMA contact symbols used to represent an off-delay timer. Notice that the arrow points in the direction the contact will move after the time delay period. The arrow indicates that the normally open contact will delay reopening, and that the normally closed contact will delay reclosing. Like on-delay timers, some off-delay timers will contain instantaneous contacts as well as timed contacts, and some will not.

Normally Open Normally Closed

Figure 7-1 NEMA standard symbols for off-delay contacts

Example Circuit 1

The circuit shown in Figure 7-2 illustrates the logic of an off-delay timer. It will be assumed that the timer has been set for a delay of five seconds. When switch S1 closes, TR coil energizes. This causes the normally open TR contacts to close immediately and turn on the lamp. When switch S1 opens, TR coil will de-energize, but the TR contacts will remain closed for five seconds before they reopen. Notice that the time delay period does not start until the coil is de-energized.

Example Circuit 2

In the second example it is assumed that timer TR has been set for a delay of ten seconds. Two

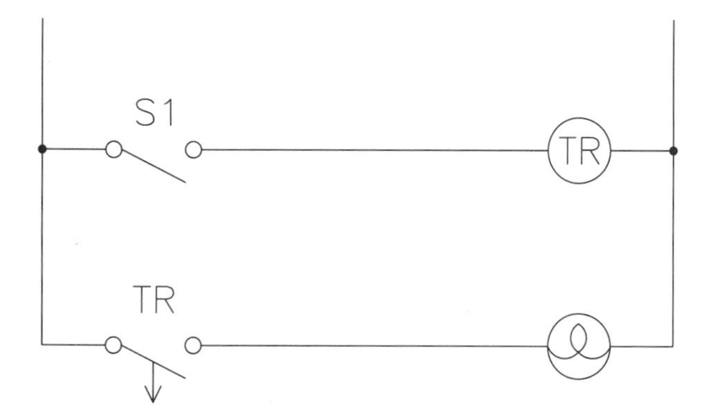

Figure 7-2 Basic operation of an off-delay timer

motors start when the start button is pressed. When the stop button is pressed, motor 1 stops operating immediately, but motor 2 continues to run for ten seconds, Figure 7-3. In this circuit the coil of the off-delay timer has been placed in parallel with motor starter 1M, permitting the action of the timer to be controlled by the first motor starter.

Example Circuit 3

Now assume that the logic of the previous circuit is to be changed so that when the start

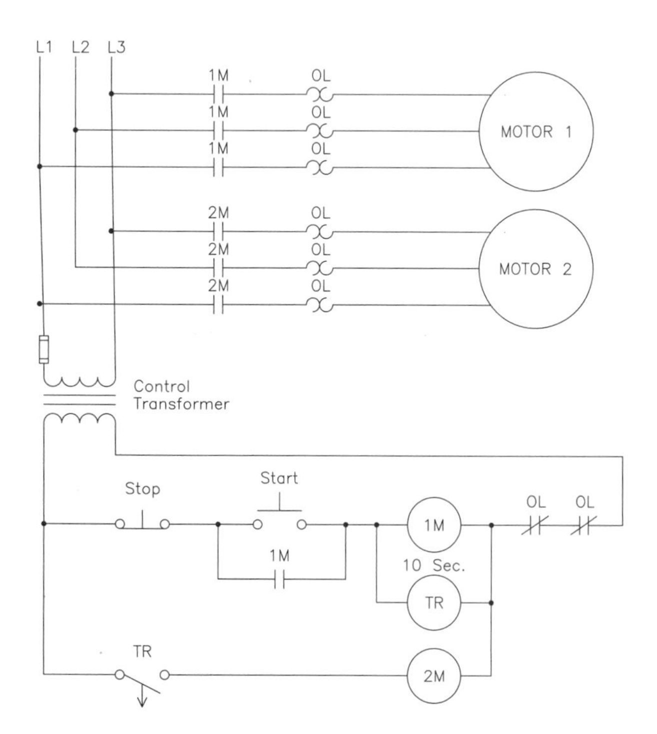

Figure 7-3 Off-delay motor circuit using pneumatic timer

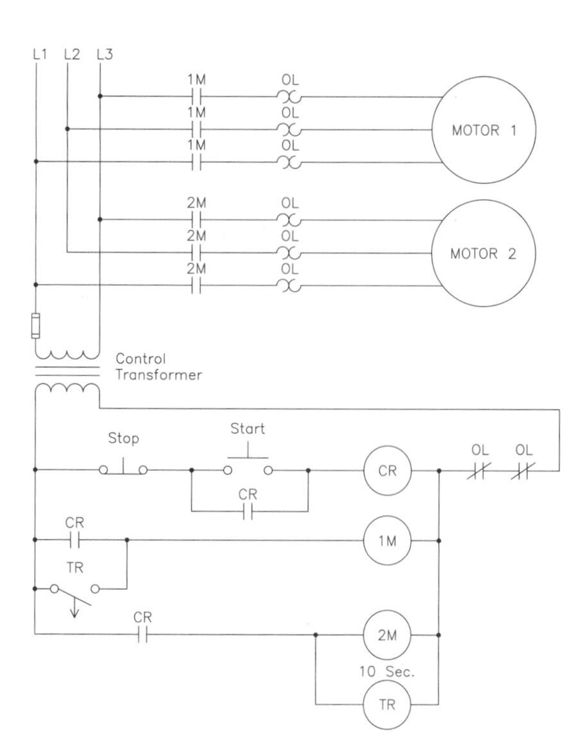

Figure 7-4 Motor 1 stops after motor 2

button is pressed, both motors still start at the same time, but when the stop button is pressed, motor 2 must stop operating immediately and motor 1 continues to run for ten seconds. In this circuit, the action of the timer must be controlled by the operation of starter 2M instead of starter 1M, Figure 7-4. In the circuit shown in Figure 7-4, a control relay is used to energize both motor starters at the same time. Notice that timer coil TR energizes at the same time as starter 2M, causing the normally open TR contacts to close around the CR contact connected in series with coil 1M.

When the stop button is pressed, coil CR de-energizes and all CR contacts open. Power is maintained to starter 1M, however, by the now closed TR contacts. When the CR contact connected in series with coils 2M and TR opens, these coils de-energize, causing motor 2 to stop operating and starting the time sequence for the off-delay timer. After a delay of ten seconds TR contacts reopen and de-energize coil 1M, stopping the operation of motor 1.

Using Electronic Timers

In the circuits shown in Figure 7-3 and Figure 7-4 it was assumed that the off-delay timers were of the pneumatic type. It is common practice to develop circuit logic assuming that the timers are of the pneumatic type. The reason for this is that the action of a pneumatic timer is controlled by the coil being energized or de-energized. The action of the timer is dependent on air pressure, not on an electric circuit. This, however, is generally not the case when using solid-state time delay relays. Solid-state timers that can be used as off-delay timers are generally designed to be plugged into an 11 pin tube socket. The pin connection for a Dayton model 6A855 timer is shown in Figure 7-5. Although this is by no means the only type of electronic timer available, it is typical of many.

Notice in Figure 7-5 that power is connected to pins 2 and 10. When this timer is used in the on-delay mode, there is no problem with the application of power because the time sequence starts when the timer is energized. When power is removed, the timer de-energizes and the contacts return to their normal state immediately.

An off-delay timer, however, does not start the timing sequence until the timer is de-energized. Because this timer depends on an electronic circuit to operate the timing mechanism, power must be connected to the timer at all times. Therefore, some means other than disconnecting the power must be used to start the timing circuit. This particular timer uses pins 5 and 6 to start the operation. The diagram in Figure 7-5 uses a start switch to illustrate this operation. When pins 5 and 6 are shorted together, it has the effect of energizing the coil of an off-delay timer and all contacts change position immediately. The timer will remain in this state as long as pins 5 and 6 are short-circuited together. When the short circuit between pins 5 and 6 is removed, it has the effect of de-energizing the coil of a pneumatic off-delay timer and the timing sequence will start. At the end of the time period the contacts will return to their normal position.

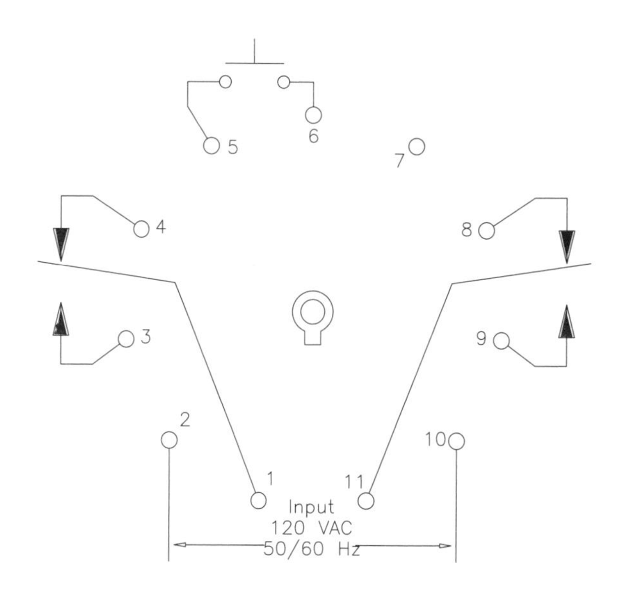

Figure 7-5 Connection diagram for a Dayton model 6A855 timer

Amending Circuit 1

The circuit in Figure 7-3 has been amended in Figure 7-6 to accommodate the use of an electronic timer. Notice in this circuit that power is connected to pins 2 and 10 of the timer at all times. Because the action of the timer in the original circuit is that the coil of the timer operates at the same time as starter coil 1M, an auxiliary contact on starter 1M will be used to control the action of timer TR. When the start button is pressed, coil 1M energizes and all 1M contacts close. This connects motor 1 to the line, the 1M contact in parallel with the start button seals the circuit, and the normally open 1M contact connected to pins 5 and 6 of the timer closes and starts the operation of the timer. When timer pins 5 and 6 become shorted, the timed contact connected in series with 2M coil closes and energizes starter 2M.

When the stop button is pressed, coil 1M de-energizes and all 1M contacts return to their normal position, stopping the operation of motor 1. When the 1M contacts connected to timer pins 5 and 6 reopen, the timing sequence of the timer begins. After a delay of ten seconds, timed contact TR reopens and disconnects starter coil 2M from the circuit. This stops the operation of motor 2.

Amending Circuit 2

Circuit 2 will be amended in much the same way as circuit 1. The timer must have power connected to it at all times, Figure 7-7. Notice in this circuit that the action of the timer is controlled by starter 2M instead of 1M. When coil 2M energizes, a set of normally open 2M contacts closes and shorts pins 5 and 6 of the timer. When coil 2M de-energizes, the 2M auxiliary contacts reopen and start the time sequence of timer TR.

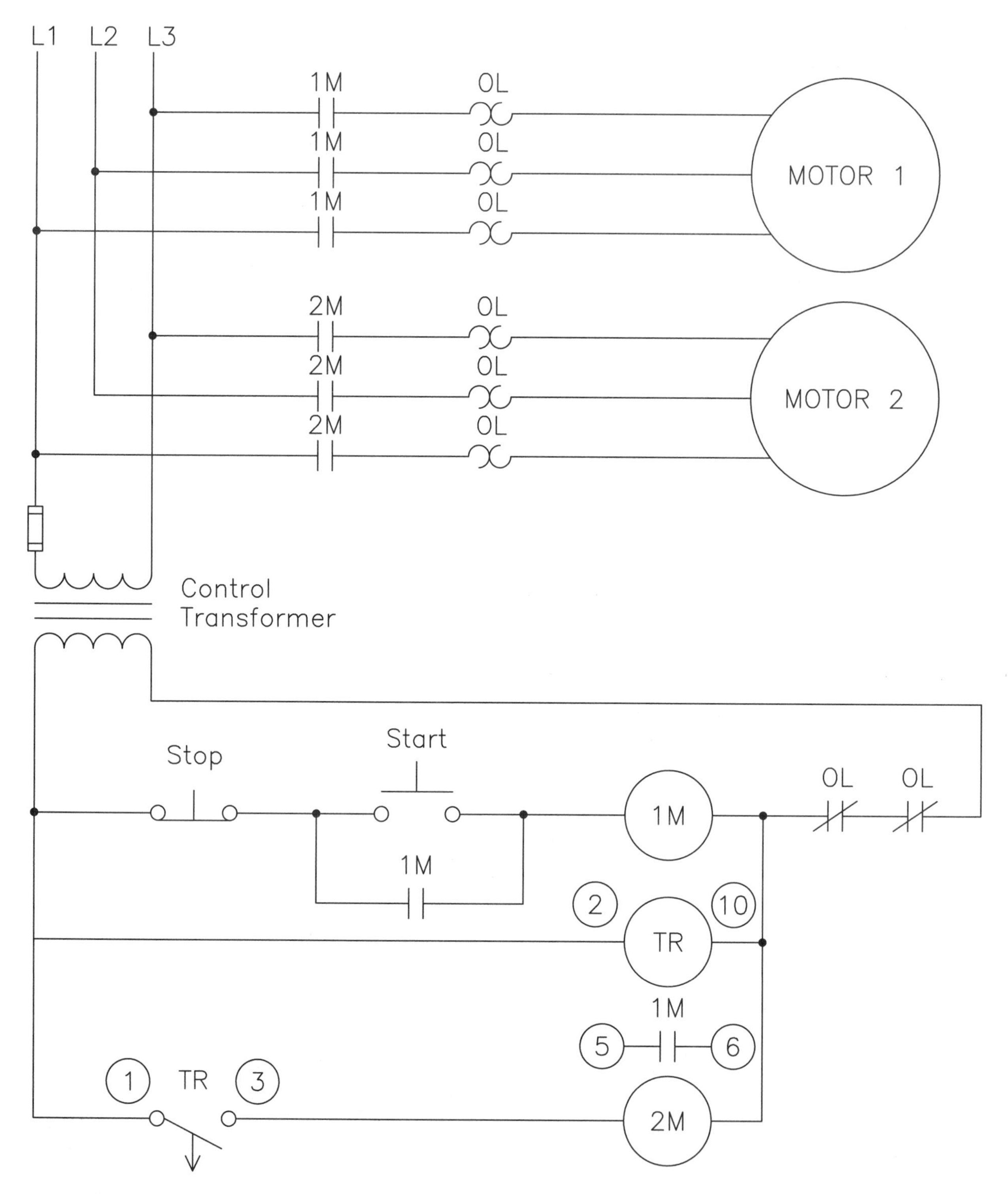

Figure 7-6 Amending the first circuit for an electronic timer

Circuit 2 assumes the use of an 11 pin control relay instead of an 8 pin. An 11 pin control relay contains three sets of contacts instead of two. Figure 7-8 shows the connection diagram for most 11 pin control relays. Notice that normally open contacts are located on pins 1 and 3, 6 and 7, and 9 and 11. The coil pins are 2 and 10. Pin numbers have been placed beside the components in Figure 7-7.

Connecting the First Circuit

1. Place wire numbers on the schematic shown in Figure 7-6.
2. Using an 11 pin tube socket, connect the control part of the circuit in Figure 7-6.
3. Set the electronic timer to operate as an off-delay timer and set the time delay for 10 seconds.
4. Plug the timer into the tube socket and turn on the power.
5. Test the control part of the circuit for proper operation.
6. If the control portion of the circuit operated properly, connect the motors or equivalent motor loads and test the entire circuit for proper operation.
7. Turn off the power and disconnect the circuit.

Connecting the Second Circuit

1. Place wire numbers on the schematic diagram shown in Figure 7-7.
2. Using two 11 pin tube sockets, connect the control part of the circuit.
3. Set the electronic timer to operate as an off-delay timer and set the time delay for 10 seconds.
4. Plug the timer and control relay into the tube sockets and turn on the power.
5. Test the control part of the circuit for proper operation.
6. If the control portion of the circuit operated properly, connect the motors or equivalent motor loads and test the entire circuit for proper operation.
7. Turn off the power and disconnect the circuit.
8. Return the components to their proper location.

Review Questions

1. Describe the operation of an off-delay timer.

2. Why is it common practice to develop circuit logic assuming that all timers are of the pneumatic type?

3. Refer to the schematic diagram shown in Figure 7-6. Assume that starter coil 2M is open. Describe the action of the circuit when the start button is pressed and when the stop button is pressed.

4. Refer to the circuit shown in Figure 7-7. Assume that when the start button is pressed, motor 1 starts operating immediately, but motor 2 does not start. When the stop button is pressed, motor 1 stops operating immediately. Which of the following could cause this condition?
   a. 1M coil is open.
   b. 2M coil is open.
   c. Timer TR is not operating.
   d. CR coil is open.

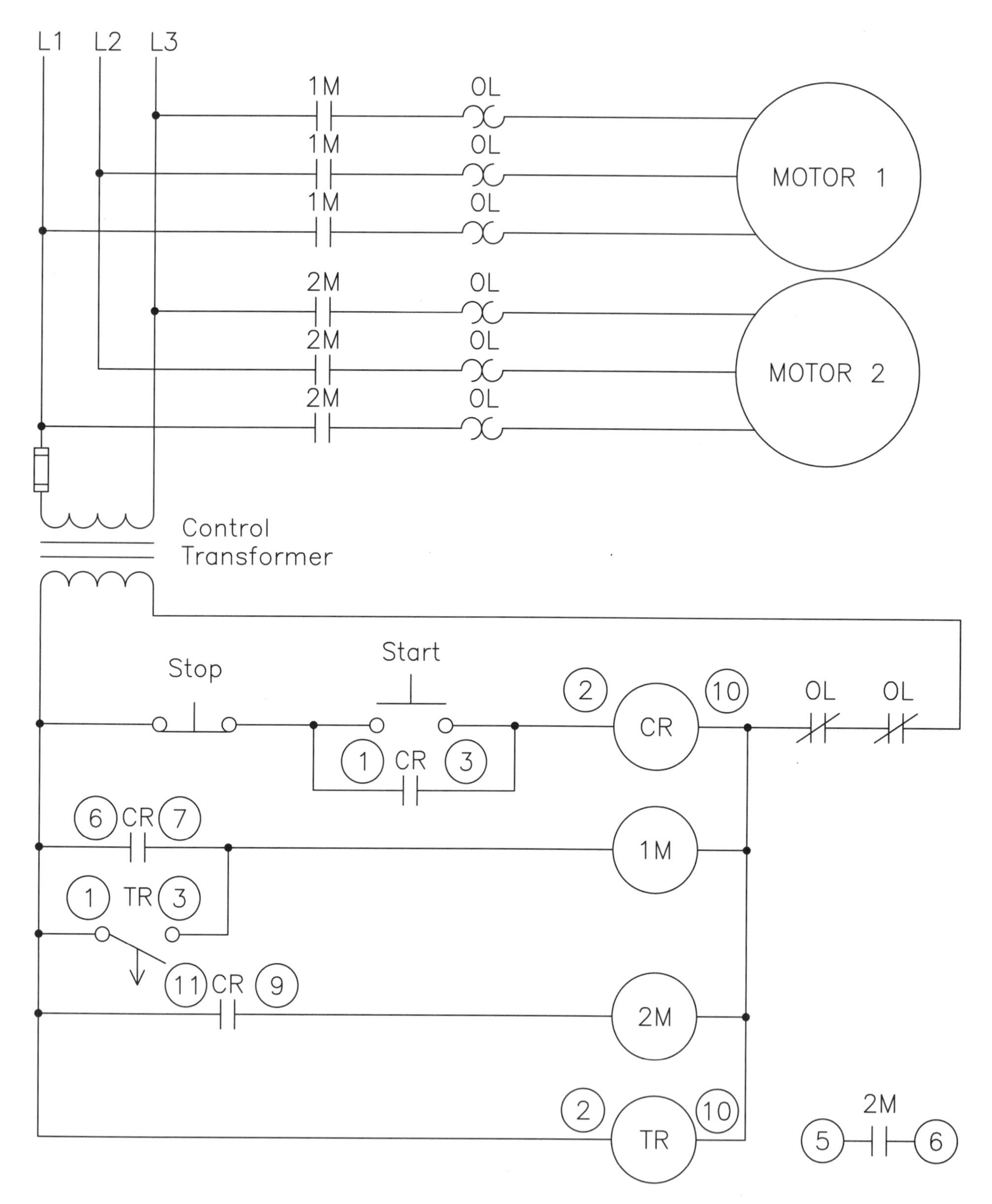

Figure 7-7 Amending circuit 2 for an electronic timer

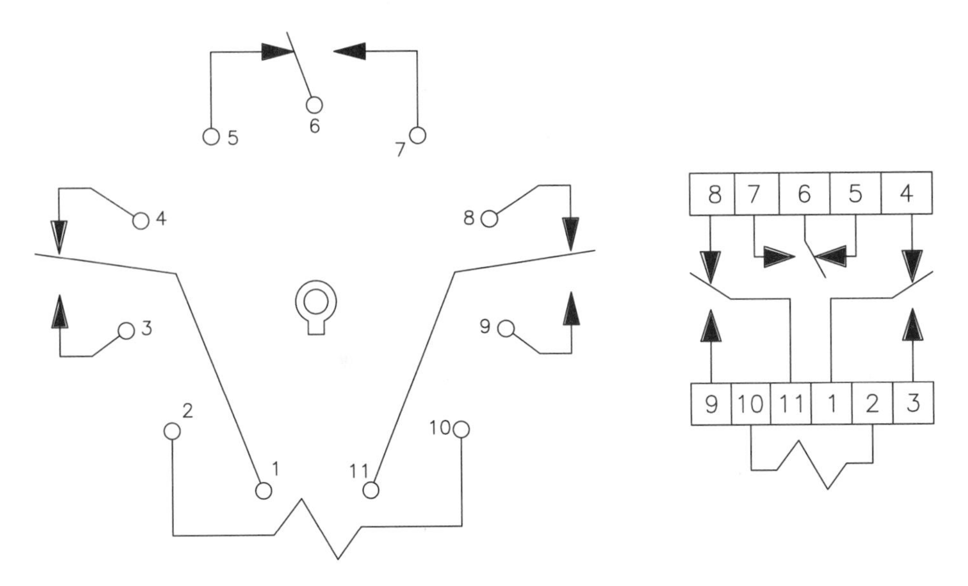

Figure 7-8 Connection diagram for an 11 pin control relay

5. Refer to the circuit shown in Figure 7-7. When the start button is pressed, both motors 1 and 2 start operating immediately. When the stop button is pressed, motor 2 stops operating immediately, but motor 1 remains running and does not turn off after the time delay period has expired. Which of the following could cause this condition?
   a. CR contacts are shorted together.
   b. 2M auxiliary contacts connected to pins 5 and 6 of the timer did not close.
   c. 2M auxiliary contacts connected to pins 5 and 6 of the timer are shorted.
   d. The stop button is shorted.

6. Refer to the circuit shown in Figure 7-7. Assume that timer TR is set for a delay of ten seconds. Now assume that timer TR is changed from an off-delay timer to an on-delay timer. Explain the operation of the circuit.

7. Using the space provided in Figure 7-9, modify the circuit in Figure 7-7 to operate as follows:
   a. When the start button is pressed, motor 1 starts running immediately. After a delay of ten seconds motor 2 begins running. Both motors remain operating until the stop button is pressed or an overload occurs.
   b. When the stop button is pressed, motor 2 stops operating immediately, but motor 1 continues to operate for a period of ten seconds before stopping.
   c. An overload on either motor will stop both motors immediately.
   d. Assume the use of electronic timers in the final design.

8. After your instructor has approved the modification, connect your circuit in the laboratory.

9. Turn on the power and test the circuit for proper operation.

10. Disconnect the power and return the components to their proper location.

Figure 7-9 Amending the circuit design

EXPERIMENT 8

Changing the Logic of an On-Delay Timer to an Off-Delay Timer

Name:	Date:	Grade:
Comments:		

Objectives

After completing this experiment, the student should be able to:

- Discuss the difference in logic between on- and off-delay timers
- Draw a schematic diagram of a circuit that will change the logic of an on-delay timer into the logic of an off-delay timer
- Connect an on-delay timer circuit that will operate with the logic of an off-delay timer

Materials Needed

- Three-phase power source
- Control transformer
- 2 ea.—Three-phase motor or equivalent motor load
- 2 ea.—Three-phase motor starters with at least two normally open and one normally closed auxiliary contacts
- 8 pin or 11 pin on-delay timer with appropriate socket
- 11 pin control relay with 11 pin socket

Some manufacturers purchase on-delay timers only. The reason for this is that most timing circuits require the logic of an on-delay timer. If it should become necessary to construct a circuit with the logic of an off-delay timer, it is a relatively simple matter to build a circuit using an on-delay timer that will operate with the same logic as an off-delay timer. A circuit of this type is shown in Figure 8-1. The basic idea is to cause the timer to start operating when a control component is turned off instead of on. Control relay CR is used to perform this function.

In the circuit shown in Figure 8-1, starter 1M is to energize immediately when switch S1 closes. When switch S1 opens, starter 1M should

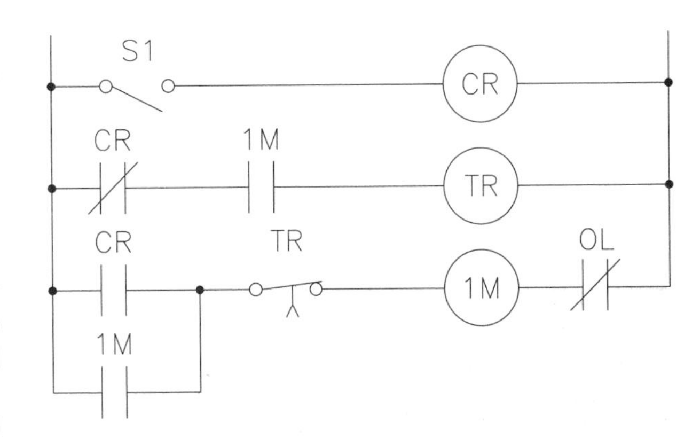

Figure 8-1 Basic circuit to change the logic of an on-delay timer into an off-delay timer

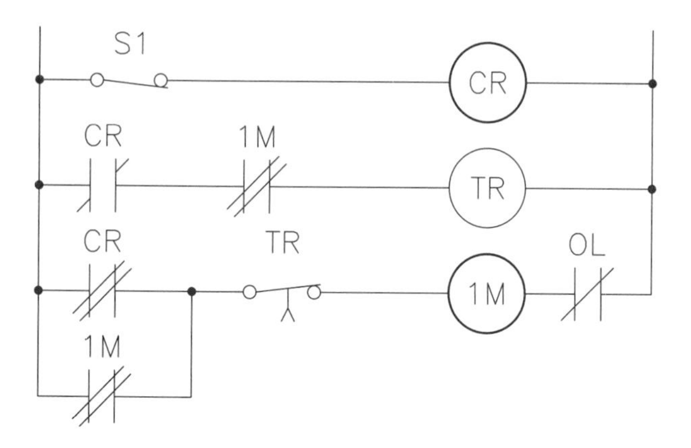

Figure 8-2 Starter 1M energizes immediately, but the timer does not start timing

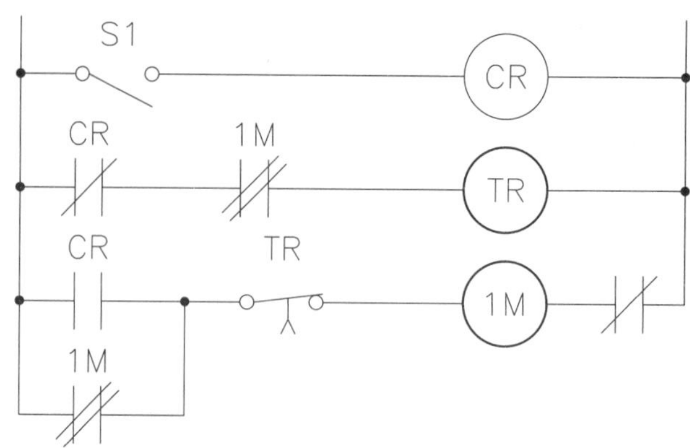

Figure 8-3 Switch S1 opens and starts the timer

remain energized for some period of time before de-energizing. This is the logic of an off-delay timer. This logic can be accomplished by using an on-delay timer and the circuit shown in Figure 8-1. When switch S1 closes, CR coil energizes and all CR contacts change position, Figure 8-2. The normally closed CR contact connected in series with TR coil opens to prevent the timer energizing. The normally open CR contact connected in series with starter coil 1M closes and energizes the coil. This causes both 1M auxiliary contacts to close. Starter 1M is now energized, but the timer has not started its time sequence.

When switch S1 is reopened, CR coil de-energizes and all CR contacts return to their normal position, Figure 8-3. When the CR contact connected in series with starter coil 1M reopens, a current path is maintained through the now closed 1M auxiliary contact connected in parallel with the open CR contact. When the CR contact connected in series with timer coil TR closed, it provided a path to coil TR and the timer began its time sequence.

At the end of the timing sequence, timed contact TR opens and de-energizes coil 1M, causing all 1M contacts to return to their normal position, Figure 8-4. The auxiliary 1M contact connected in series with timer coil TR opens, and de-energizes coil TR. This causes contact TR to reclose, and the circuit is back to the beginning state shown in Figure 8-1.

Changing an Existing Schematic

The circuit shown in Figure 8-5 is an off-delay timer circuit for the control of two motors. It is assumed that the timer used in this circuit is a pneumatic timer. This circuit was discussed in the previous unit. Both motors start when the start button is pressed. When the stop button is pressed, motor 2 stops operating immediately, but motor 1 continues to operate for a period of ten seconds. Now assume that it is necessary to change the circuit logic to permit an on-delay timer to be used.

Notice in the circuit in Figure 8-5 that timer coil TR is energized or de-energized at the same time as starter coil 2M. In the amended circuit, starter 2M will control the starting of on-delay timer TR, Figure 8-6. A set of 1M auxiliary contacts prevents coil TR from being energized until starter 1M has been energized. To understand the operation of the circuit, trace the logic through each step of operation. Assume that the start button is pushed and coil CR energizes. This causes all CR contacts to close and connect

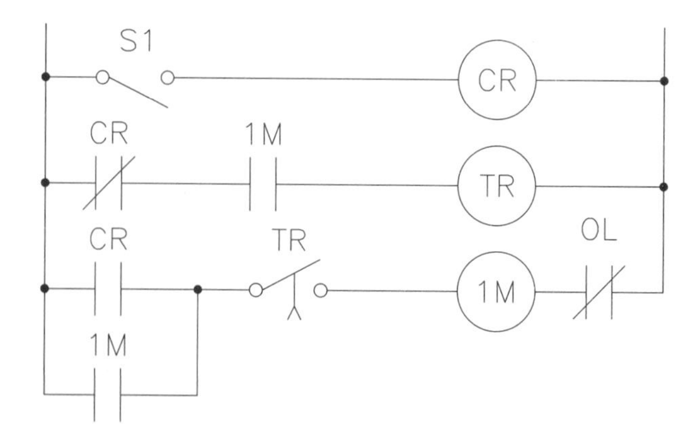

Figure 8-4 Starter 1M de-energizes when timer contact TR opens

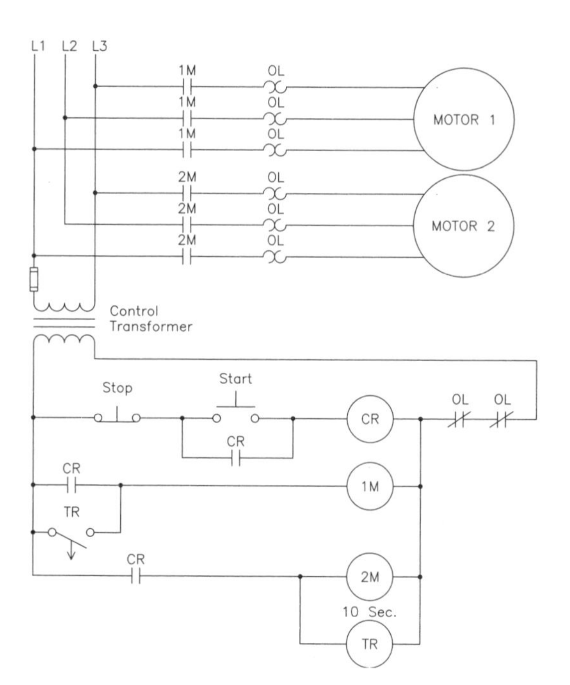

Figure 8-5 Off-delay timer circuit using a pneumatic timer

starters 1M and 2M to the line, Figure 8-7. Both 1M auxiliary contacts close, but the normally closed 2M auxiliary contact connected in series with TR coil opens and prevents it from starting its time sequence.

When the stop button is pressed, CR coil de-energizes and all CR contacts return to their normal position, Figure 8-8. Motor starter 1M remains energized because of the closed 1M auxiliary contact connected in parallel with the CR contact. When starter 2M de-energizes, the normally closed auxiliary contact connected in series with timer coil TR recloses and on-delay timer TR begins its timing sequence.

After a delay of ten seconds, timed contact TR opens and disconnects starter coil 1M from the line, Figure 8-9. This stops the operation of motor 1 and returns all 1M auxiliary contacts to their normal position. When timer TR de-energizes, timed contact TR returns to its normally closed position and the circuit is back to its original state shown in Figure 8-6.

Connecting the Circuit

1. Using the circuit shown in Figure 8-6, place pin numbers beside the control components that mount into tube sockets. These components will probably be the control relay and the timer. Be sure to place pin numbers beside contacts as well as coils. Circle the pin numbers to distinguish them from wire numbers.
2. Place wire numbers on the schematic diagram.
3. Connect the control part of the circuit.
4. Turn on the power and test the circuit for proper operation.
5. If the control part of the circuit operates properly, turn off the power and connect the motors or equivalent motor loads.
6. Turn on the power and test the entire circuit for proper operation.
7. Turn off the power and disconnect the circuit.

Review Questions

1. Why do some companies purchase only on-delay timers?

2. Refer to the circuit shown in Figure 8-10. This circuit assumes the use of a pneumatic off-delay timer. It is also assumed that the timer is set for a delay of ten seconds. Describe the operation of this circuit when the start push button is pressed.

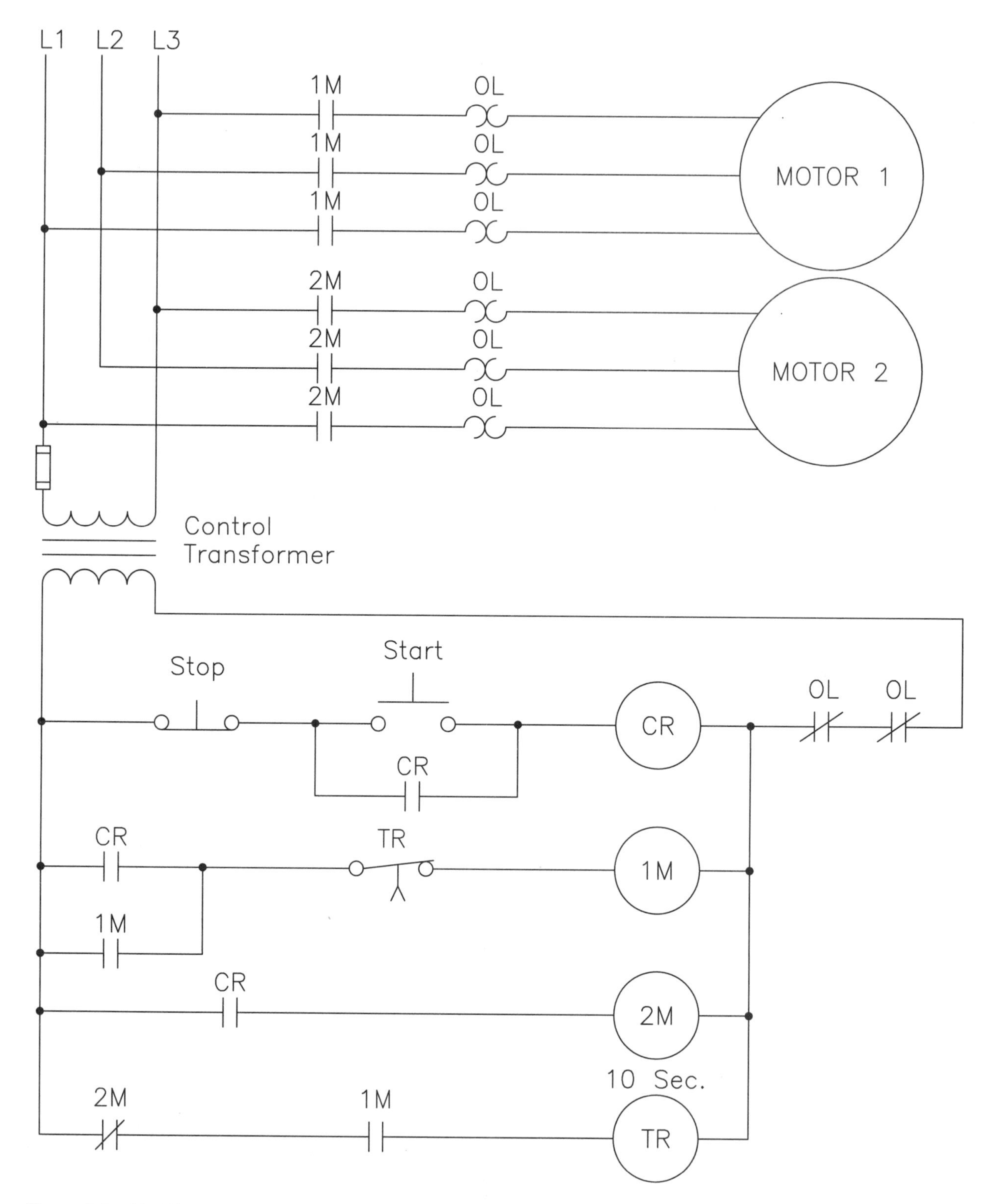

Figure 8-6 Modifying the circuit for an on-delay timer

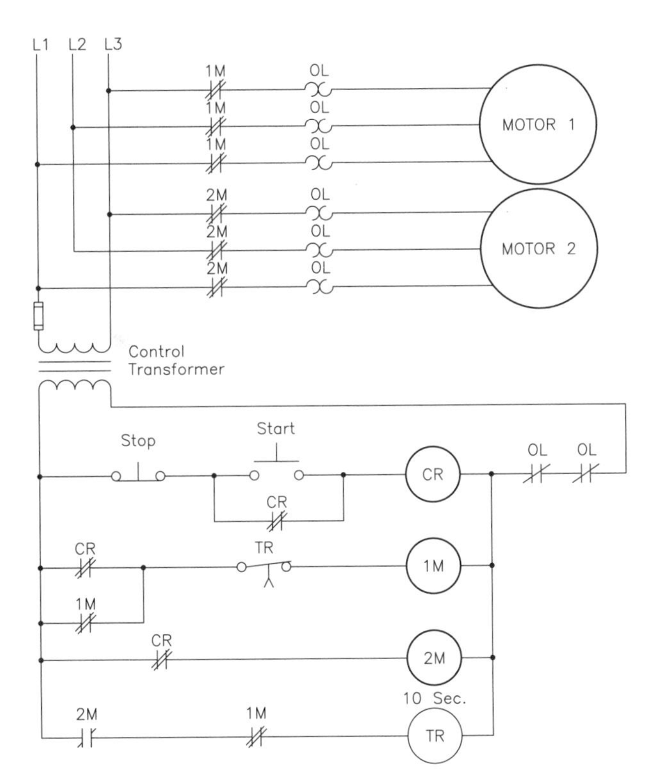

Figure 8-7 Both motors start at the same time

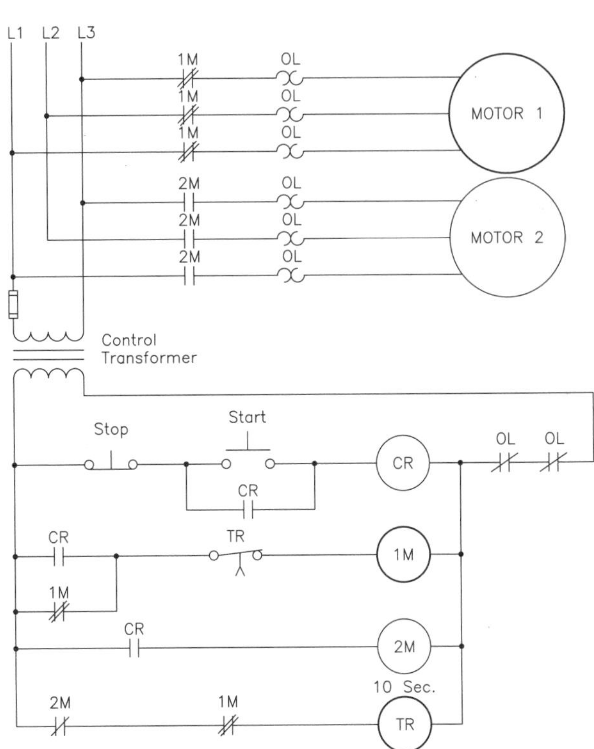

Figure 8-8 Starter 2M de-energizes; timer TR starts its time sequence

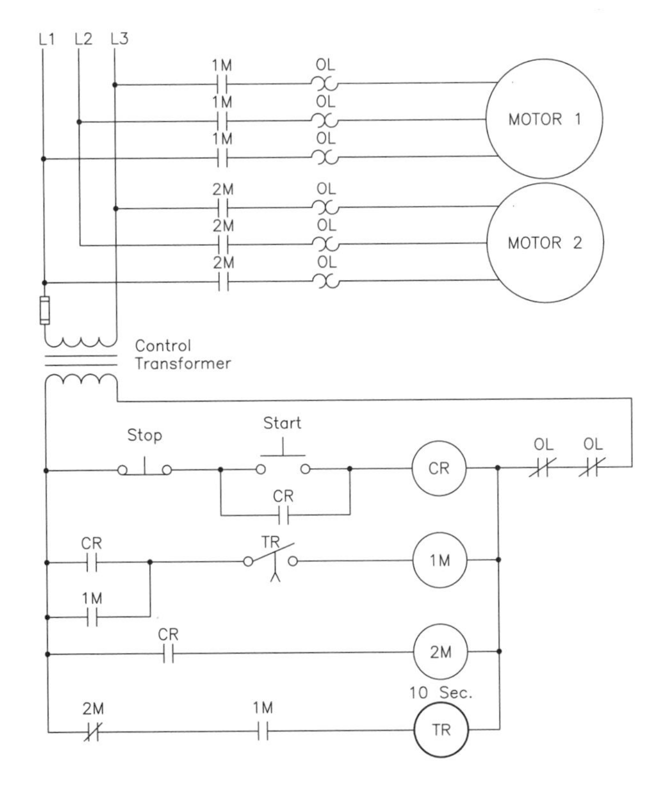

Figure 8-9 Timed contact de-energizes starter 1M

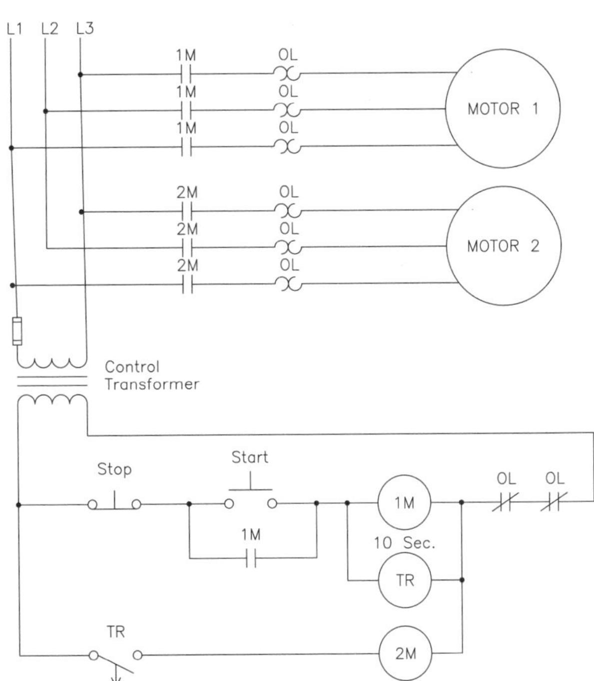

Figure 8-10 Motor 1 stops operating before motor 2

3. Assume that the circuit in Figure 8-10 is in operation. Describe the action of the circuit when the stop button is pressed.

4. The circuit shown in Figure 8-10 employs a pneumatic off-delay timer. Redraw the circuit in the space provided in Figure 8-11 to use an electronic on-delay timer. Make certain that the logic of the circuit is the same.

5. After your instructor has approved the redrawn circuit, connect the circuit in the laboratory.

6. Turn on the power and test the circuit for proper operation.

7. Turn off the power and return the components to their proper place.

Figure 8-11 Circuit redesign

EXPERIMENT 9

Designing a Printing Press Circuit

Name: Date: Grade:

Comments:

Objectives

After completing this experiment, the student should be able to:

- Describe a step-by-step procedure for designing a motor control circuit
- Design a basic control circuit
- Connect the completed circuit in the laboratory

Materials Needed

- Three-phase power supply
- Three-phase motor starter
- 8 or 11 pin on-delay relay with appropriate socket
- Three-phase motor or equivalent motor load
- Pilot light
- Buzzer or simulated load
- Control transformer
- 8 pin control relay and 8 pin socket

In this experiment a circuit for a large printing press will be designed in a step-by-step procedure. The owner of a printing company has the following concern when starting a large printing press:

The printing press is very large, and the surrounding noise level is high. There is a danger that when the press starts, a person unseen by the operator may have his or her hands in the press. To prevent an accident, I would like to install a circuit that sounds an alarm and flashes a light for ten seconds before the press actually starts. This would give the person time to get clear of the machine before it starts.

To begin the design procedure, list the requirements of the circuit. List not only the concerns of the owner, but also any electrical or safety requirements that the owner may not be aware of. Understand that the owner is probably not an electrical technician and does not know all the electrical requirements of a motor control circuit.

1. There must be a start and stop push button control.
2. When the start button is pressed, a warning light and buzzer turn on.
3. After a delay of ten seconds, the warning light and buzzer turn off and the press motor starts.
4. The press motor should be overload protected.
5. When the stop push button is pressed, the circuit will de-energize even if the motor has not started.

To begin design of the circuit, fulfill the first requirement of the logic: When the start button is pressed, a warning light and buzzer turn on for a period of ten seconds. This first part of the circuit can be satisfied with the circuit shown in Figure

9-1. In this example, a timer is used because the warning light and buzzer are to remain on for only ten seconds. Because the warning light and buzzer are to turn on immediately when the start button is pressed, a normally closed timed contact is used. This circuit also assumes that the timer contains an instantaneous contact that is used to hold the circuit in after the start button is released.

The next part of the logic states that after a delay of ten seconds the warning light and buzzer are to turn off and the press motor is to start. As the present circuit is shown in Figure 9-1, when the start button is pressed, TR coil will energize. This causes the normally open instantaneous TR contacts to close and hold TR coil in the circuit when the start button is released. At the same time, timer TR starts its timing sequence. After a delay of ten seconds the normally closed TR timed contact connected in series with the warning light and buzzer will open and disconnect them from the circuit.

The only remaining circuit logic is to start the motor after the warning light and buzzer have turned off. This can be accomplished with a normally open timed contact controlled by timer TR, Figure 9-2. At the end of the timing sequence, the normally closed TR contact will open and disconnect the warning light and buzzer. At the same time, the normally open TR timed contact will close and energize the coil of M starter. The normally closed overload contact connected in series with the rest of the circuit will de-energize the entire circuit in the event of motor overload.

Now that the logic of the control circuit has been completed, the motor load can be added as shown in Figure 9-3.

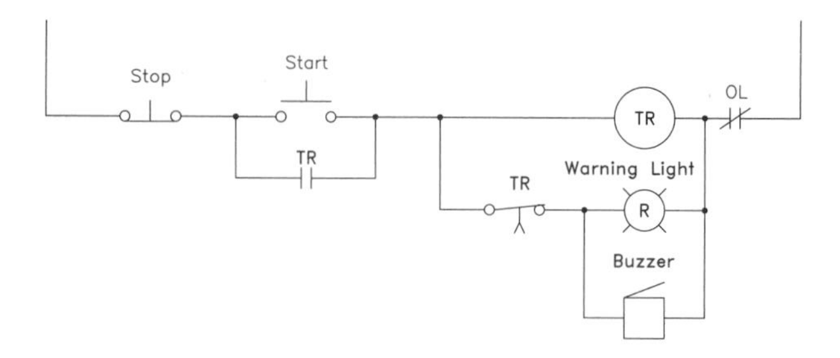

Figure 9-1 First step in the circuit design

Addressing a Potential Problem

The completed circuit shown in Figure 9-3 assumes the use of a timer that contains both timed and instantaneous contacts. This contact arrangement is common for certain types of timers such as pneumatic and some clock timers, but most electronic timers do not contain instantaneous contacts. If this is the case, a control relay can be added to supply the needed instantaneous contact by connecting the coil of the control relay in parallel with the coil of TR timer, Figure 9-4.

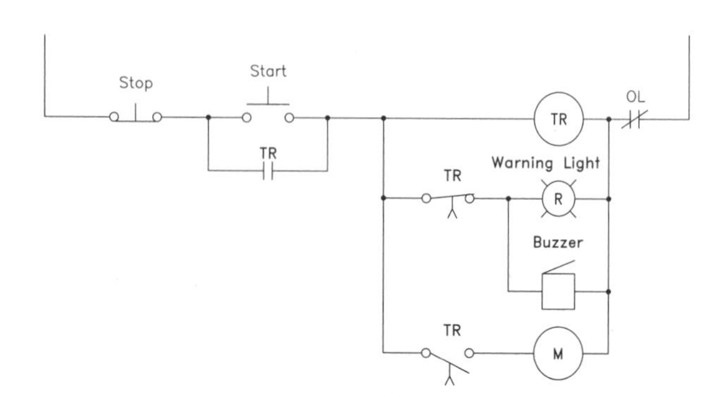

Figure 9-2 Completing the circuit logic

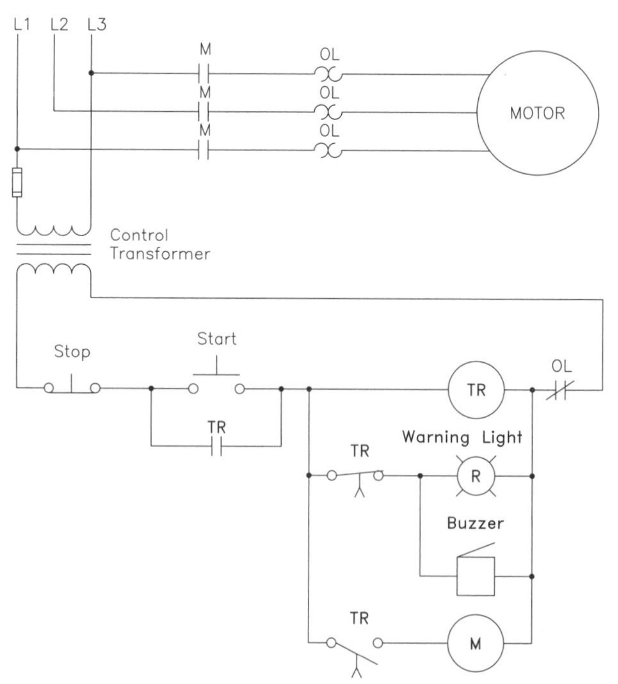

Figure 9-3 The complete circuit

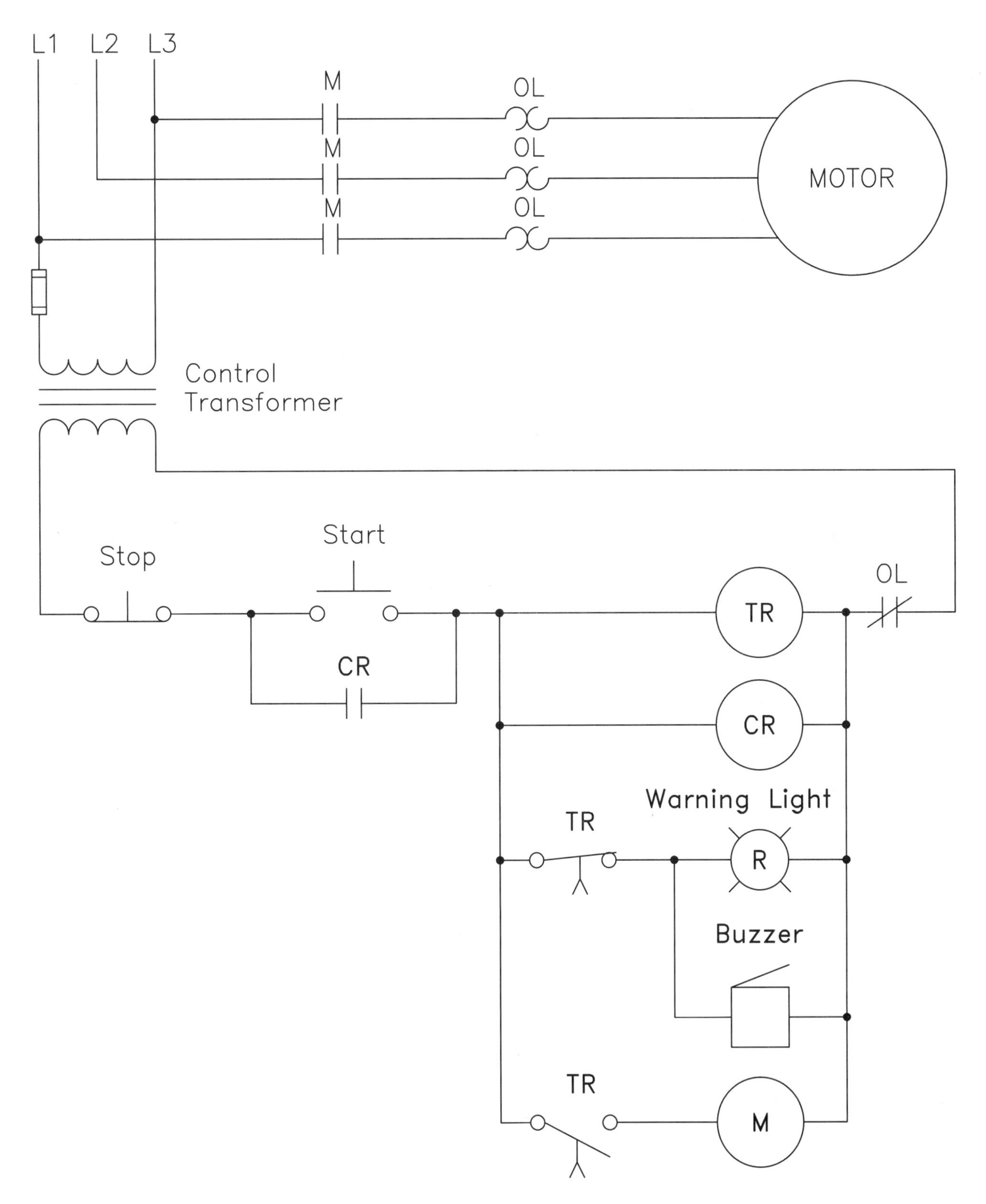

Figure 9-4 Adding a control relay

Connecting the Circuit

1. It will be assumed that the timer in this circuit is the electronic type. Therefore, it will be assumed that a control relay will be used to provide the normally open holding contacts. Assuming the use of an electronic on-delay timer and an 8 pin control relay, place pin numbers beside the components of the timer and control relay shown in Figure 9-4. Circle the numbers to distinguish them from wire numbers.
2. Place wire numbers beside the components in Figure 9-4.
3. Connect the control portion of the circuit. (***Note:*** It may be necessary to use a pilot light for the buzzer if one is not available.)
4. Turn on the power and test the control part of the circuit for proper operation.
5. Turn the power off.
6. If the control part of the circuit operated properly, connect the motor or simulated motor load to the circuit.
7. Turn on the power and test the entire circuit for proper operation.
8. Turn off the power and return the components to their proper location.

Review Questions

1. What should be the first step when beginning the design of a control circuit?

2. Why is it sometimes necessary to connect the coil of a control relay in parallel with the coil of a timer?

3. Refer to the circuit shown in Figure 9-3. Assume that the on-delay timer is replaced with an off-delay timer. Describe the action of the circuit when the start button is pressed.

4. Describe the operation of the circuit when the stop button is pressed. Assume that the circuit is running with an off-delay timer as described in review question 3.

5. Refer to the circuit shown in Figure 9-4. Assume that the owner decides to change the logic of the circuit as follows:

   When the operator presses the start button, a warning light and buzzer turn on for a period of ten seconds. During this ten seconds, the operator must continue to hold down the start button. If the start button should be released, the timing sequence will stop and the motor will not start. At the end of ten seconds, provided that the operator continues to hold the start button down, the warning light and buzzer will turn off and the motor will start. When the motor starts, the operator can release the start button and the press will continue to run.

   Amend the circuit in Figure 9-4 to meet the requirement.

EXPERIMENT 10

Sequence Starting and Stopping for Three Motors

Name: Date: Grade:

Comments:

Objectives

After completing this experiment, the student should be able to:

- Discuss the step-by-step procedure for designing a circuit
- Change a circuit designed with pneumatic timers into a circuit to use electronic timers
- Connect the circuit in the laboratory
- Troubleshoot the circuit

Materials Needed

- Three-phase power supply
- Control transformer
- 2 ea.—8 pin control relays and 8 pin sockets
- 3 ea.—Three-phase motor starters
- 4 ea.—Electronic timers (Dayton model 6A855 or equivalent) and 11 pin sockets
- 3 ea.—Three-phase motors or equivalent motor loads

In this experiment a circuit will be designed and connected. The requirements of the circuit are as follows:

1. Three motors are to start in sequence from motor 1 to motor 3.
2. There is to be a time delay of three seconds between the starting of each motor.
3. When the stop button is pressed, the motors are to stop in sequence from motor 3 to motor 1.
4. There is to be a time delay of three seconds between the stopping of each motor.
5. An overload on any motor will stop all motors.

When designing a control circuit, satisfy one requirement at a time. This may, at times, lead to an unforeseen deadend, but don't let these deadends concern you. When they happen, backup and redesign around them. In this example, the first part of the circuit is to start three motors in sequence from motor 1 to motor 3 with a three-second delay between the starting of each motor. This is also the time to satisfy the requirement that an overload on any motor will stop all motors. The first part of the circuit can be satisfied by the circuit shown in Figure 10-1. (***Note:*** In this experiment the motor connections will not be shown because of space limitations. It is to be assumed that the motor starters are controlling three-phase motors. It is also assumed that all timers are set for a delay of three seconds.)

When the start button is pressed, coils 1M and TR1 energize. Starter 1M starts motor 1

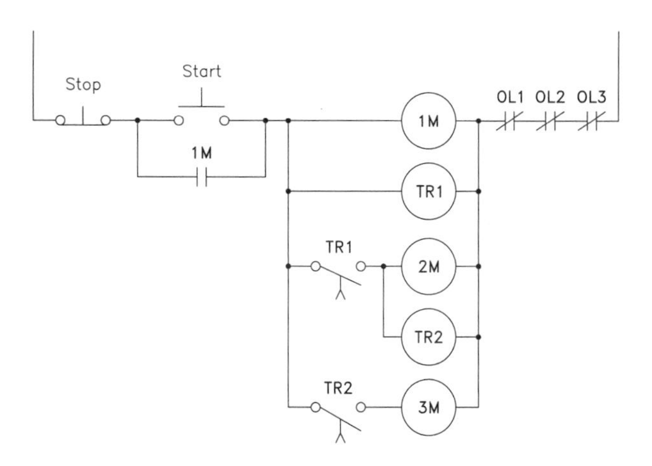

Figure 10-1 The motors start in sequence from 1 to 3

immediately, and timer TR1 starts its time sequence of three seconds. After a delay of three seconds, timed contact TR1 closes and energizes coils 2M and TR2. Starter 2M starts motor 2 and timer TR2 begins its three-second timing sequence. After a delay of three seconds, timed contact TR2 closes and energizes motor 3. The motors have been started in sequence from 1 to 3 with a delay of three seconds between the starting of each motor. This satisfies the first part of the circuit logic.

The next requirement is that the circuit stop in sequence from motor 3 to motor 1. To fulfill this requirement, power must be maintained to starters 2M and 1M after the stop button has been pushed. In the circuit shown in Figure 10-1, this is not possible. Because all coils are connected after the M auxiliary holding contact, power will be disconnected from all coils when the stop button is pressed and the holding contact opens. This circuit has proven to be a deadend. There is no way to fulfill the second requirement with the circuit connected in this manner. Therefore, the circuit must be amended in such a manner that it will not only start in sequence from motor 1 to motor 3 with a three-second time delay between the starting of each motor, but it will also be able to maintain power after the start button is pressed. This amendment is shown in Figure 10-2.

To modify the circuit so that power can be maintained to coils 2M and 1M, a control relay has been added to the circuit. Contact $1CR_2$ prevents power from being applied to coils 1M and TR1 until the start button is pressed.

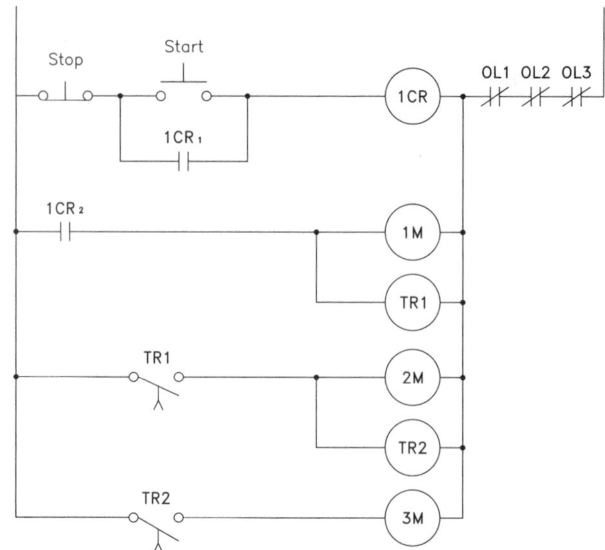

Figure 10-2 A control relay is added to the circuit

Designing the Second Part of the Circuit

The second part of the circuit states that the motors must stop in sequence from motor 3 to motor 1. Do not try to solve all the logic at once. Solve each problem as it arises. The first problem is to stop motor 3. In the circuit shown in Figure 10-2, when the stop button is pressed, coil 1CR will de-energize. This will cause contact $1CR_2$ to open and de-energize coils 1M and TR1. Contact TR1 will open immediately and de-energize coils 2M and TR2, causing contact TR2 to open immediately and de-energize coil 3M. Notice that coil 3M does de-energize when the stop button is pressed, but so does everything else. The circuit requirement states that there is to be a three-second time delay between the stopping of motor 3 and motor 2. Therefore, an off-delay timer will be added to maintain connection to coil 2M after coil 3M has de-energized, Figure 10-3.

The same basic problem exists with motor 1. In the present circuit, motor 1 will turn off immediately when the stop button is pressed. To help satisfy the second part of the problem, another off-delay relay must be added to maintain a circuit to motor 1 for a period of three seconds after motor 2 has turned off. This addition is shown in Figure 10-4.

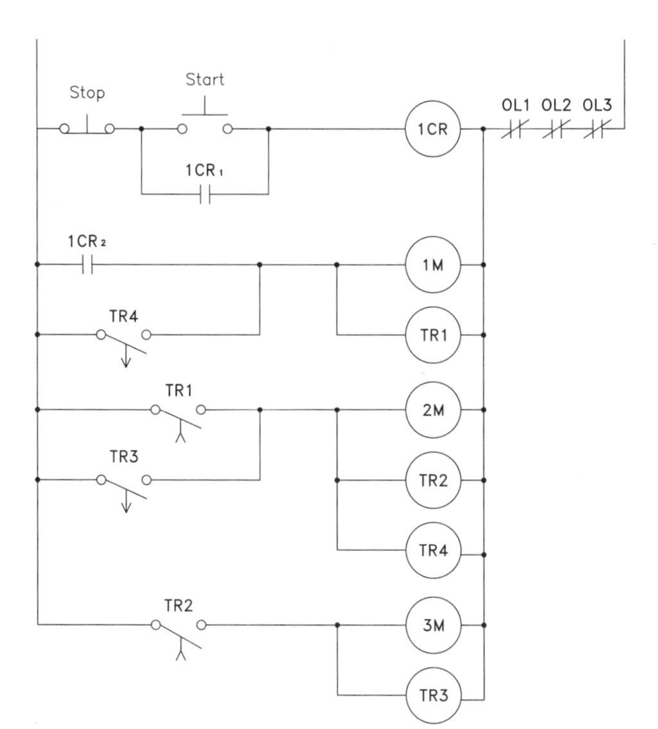

Figure 10-3 Timer TR3 prevents motor 2 from stopping

Figure 10-4 Off-delay timer TR4 prevents motor 1 from stopping

Motors 2 and 1 will now continue to operate after the stop button is pressed, but so will motor 3. In the present design, none of the motors will turn off when the stop button is pressed. To understand this condition, trace the logic step by step. When the start button is pressed, coil 1CR energizes and closes all 1CR contacts. When contact $1CR_2$ closes, coils 1M and TR1 energize. After a period of three seconds, timed contact TR1 closes and energizes coils 2M, TR2, and TR4. Timed contact TR4 closes immediately to bypass contact $1CR_2$. After a delay of three seconds, timed contact TR2 closes and energizes coils 3M and TR3. Timed contact TR3 closes immediately and bypasses contact TR1. When the stop button is pressed, coil 1CR de-energizes and all 1CR contacts open, but a circuit is maintained to coils 1M and TR1 by contact TR4. This prevents timed contact TR1 from opening to de-energize coils 2M, TR2, and TR4, which in turn prevents timed contact TR2 from opening to de-energize coils 3M and TR3. To overcome this problem, two more contacts controlled by relay 1CR will be added to the circuit, Figure 10-5. The circuit will now operate in accord with all the stated requirements.

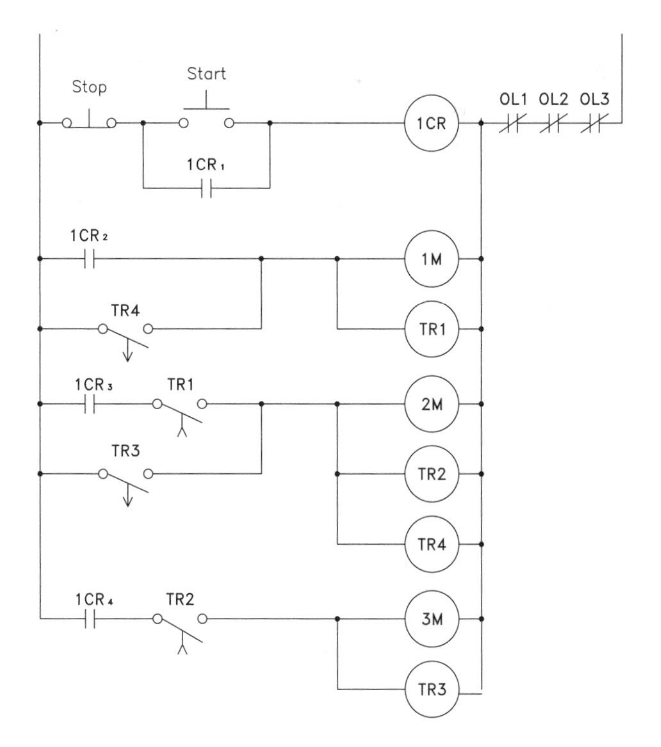

Figure 10-5 Control relay contacts are added to permit the circuit turn off

Modifying the Circuit

The circuit in Figure 10-5 was designed with the assumption that all the timers are of the pneumatic type. When this circuit is connected in the laboratory, 8 pin control relays and electronic timers will be used. The circuit will be amended to accommodate these components. The first change to be made concerns the control relays. Notice that the circuit requires the use of four normally open contacts controlled by coil 1CR. Since 8 pin control relays have only two normally open contacts, it will be necessary to add a second control relay, 2CR. The coil of relay 2CR will be connected in parallel with 1CR, which will permit both to operate at the same time, Figure 10-6.

Timers TR1 and TR2 are on-delay timers and do not require an adjustment in the circuit logic to operate. Timers TR3 and TR4, however, are off-delay timers and do require changing the circuit. The coils must be connected to power at all times. Assuming the use of a Dayton timer model 6A855, power would connect to pins 2 and 10. Starter 3M will be used to control the action of timer TR3 by connecting a 3M normally open auxiliary contact to pins 5 and 6 of timer TR3, Figure 10-7. Starter 2M will control the action of timer TR4 by connecting a 2M normally open auxiliary contact to pins 5 and 6 of that timer. The circuit is now complete and ready for connection in the laboratory.

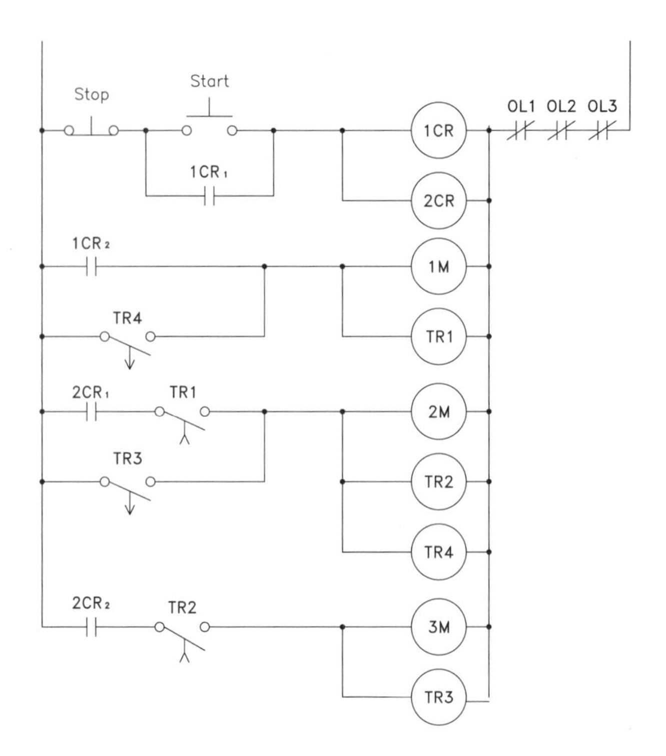

Figure 10-6 Adding a control relay to the circuit

Connecting the Circuit

1. Using the circuit shown in Figure 10-7, place pin numbers beside the proper components. Circle the pin numbers to distinguish them from wire numbers.
2. Place wire numbers on the schematic.
3. Connect the control circuit in the laboratory.
4. Turn on the power and test the circuit for proper operation.
5. Turn off the power and connect the motor loads to starters 1M, 2M, and 3M.
6. Turn on the power and test the complete circuit.
7. Turn off the power.
8. Disconnect the circuit and return the components to their proper place.

Review Questions

Refer to the circuit in Figure 10-7 to answer the following questions. It is assumed that all timers are set for a delay of three seconds.

1. When the start button is pressed, motor 1 starts operating immediately. Three seconds later motor 2 starts, but motor 3 never starts. When the stop button is pressed, motor 2 stops operating immediately. After a delay of three seconds, motor 1 stops running. Which of the following could not cause this condition?
   a. TR3 coil is open.
   b. 3M coil is open.
   c. TR2 coil is open.
   d. 2CR coil is open.

2. When the start button is pressed, motor 1 starts operating immediately. Motor 2 does not start operating after three seconds, but after a delay of six seconds motor 3 starts operating. When the stop button is pushed, motors 3 and 1 stop operating immediately. Which of the following could cause this condition?
   a. 2CR coil is open.
   b. TR1 coil is open.
   c. TR3 coil is open.
   d. 2M coil is open.
3. When the start button is pressed, all three motors start normally with a three-second delay between the starting of each motor. When the stop button is pressed, motor 3 stops operating immediately. After a delay of three seconds, both motors 2 and 1 stop operating at the same time. Which of the following could cause this problem?
   a. Timer TR1 is defective.
   b. Timer TR2 is defective.
   c. Timer TR3 is defective.
   d. Timer TR4 is defective.
4. When the start button is pressed, nothing happens. None of the motors starts. Which of the following could **not** cause this problem?
   a. Overload contact OL1 is open.
   b. 1CR relay coil is open.
   c. 2CR relay coil is open.
   d. The stop button is open.
5. When the start button is pressed, motor 1 does not start, but after a delay of three seconds motor 2 starts, and three seconds later motor 3 starts. When the stop button is pressed, motor 3 stops running immediately and after a delay of three seconds motor 2 stops running. Which of the following could cause this problem?
   a. Starter coil 1M is open.
   b. TR1 timer coil is open.
   c. Timer TR4 is defective.
   d. 1CR coil is open.

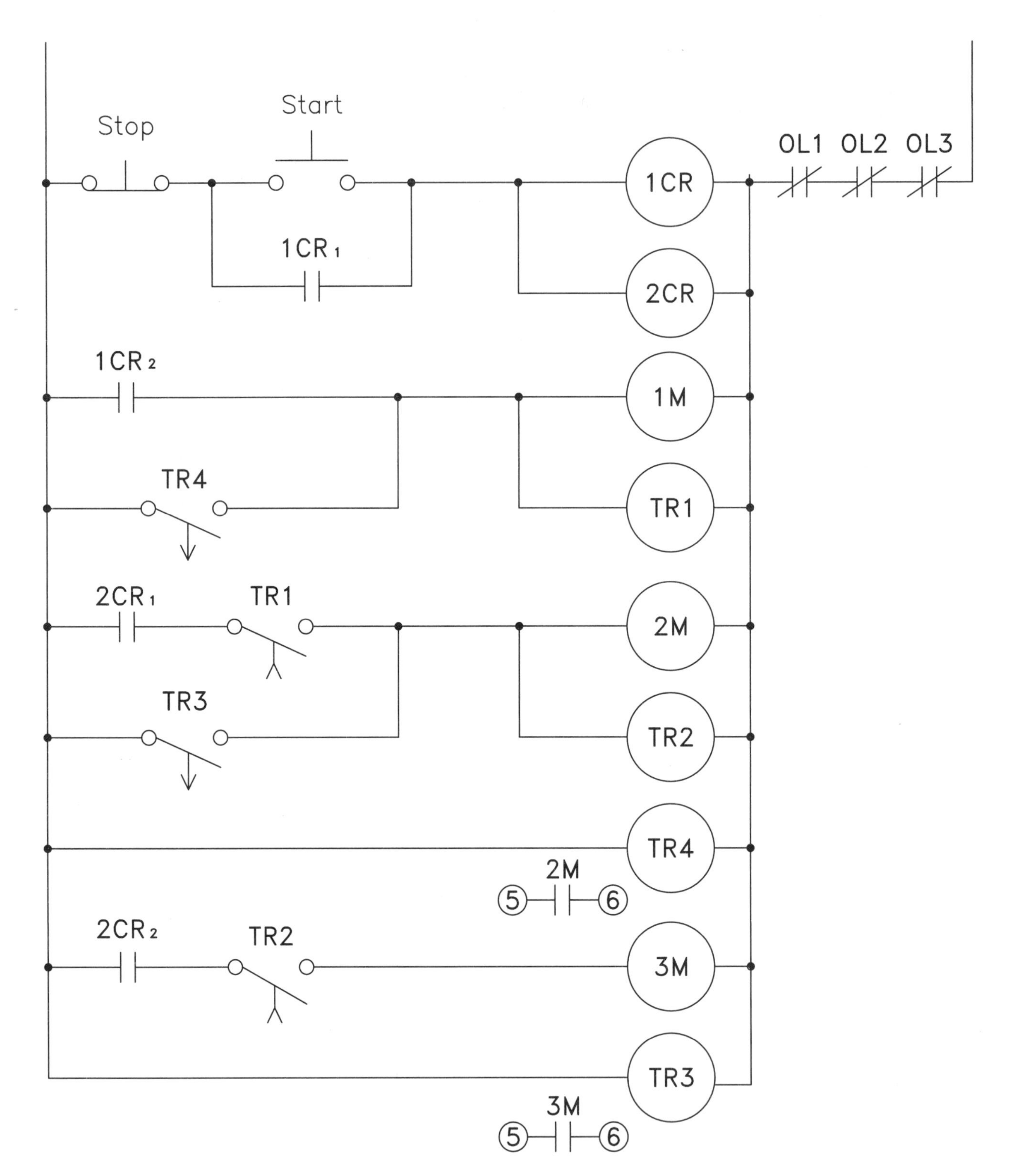

Figure 10-7 Changing pneumatic timers for electronic timers

EXPERIMENT 11

Hydraulic Press Control Circuit

Name: Date: Grade:

Comments:

Objectives

After completing this experiment, the student should be able to:

- Discuss the operation of this hydraulic press control circuit
- Connect the circuit in the laboratory
- Operate the circuit using toggle switches to simulate limit and pressure switches

Materials Needed

- Three-phase power supply
- Control transformer
- Three-phase motor starter with at least two normally open auxiliary contacts
- 5 ea.—Double acting push buttons (N.O./N.C. on each button)
- Pilot light
- 3 ea.—Toggle switches that can be used to simulate two limit switches and one pressure switch
- One three-phase motor or equivalent motor load
- 2 ea.—Solenoid coils or lamps to simulate solenoid coils
- 3 ea.—Control relays with three sets of contacts (11 pin) and 11 pin sockets
- 3 ea.—Control relays with two sets of contacts (8 pin) and 8 pin sockets

---

The next circuit to be discussed is a control for a large hydraulic press, Figure 11-1. In this circuit, a hydraulic pump must be started before the press can operate. Pressure switch PS closes when there is sufficient hydraulic pressure to operate the press. If switch PS should open, it will stop the operation of the circuit. A green pilot light is used to tell the operator that there is enough pressure to operate the press.

Two run push buttons are located far enough apart so that both of the operator's hands must be used to cause the press to cycle. This is to prevent the operator from getting his hands in the press when it is operating. Limit switches UPLS and DNLS are used to determine when the press is at the bottom of its downstroke and when it is at the top of its upstroke. In the event one or both of the run push buttons are released during the cycle, a reset button can be used to reset the press to its top position. The up solenoid causes the press to travel upward when it is energized, and the down solenoid causes the press to travel downward when it is energized.

To understand the operation of this circuit, assume that the press is in the up position. Notice that limit switch UPLS is shown normally open

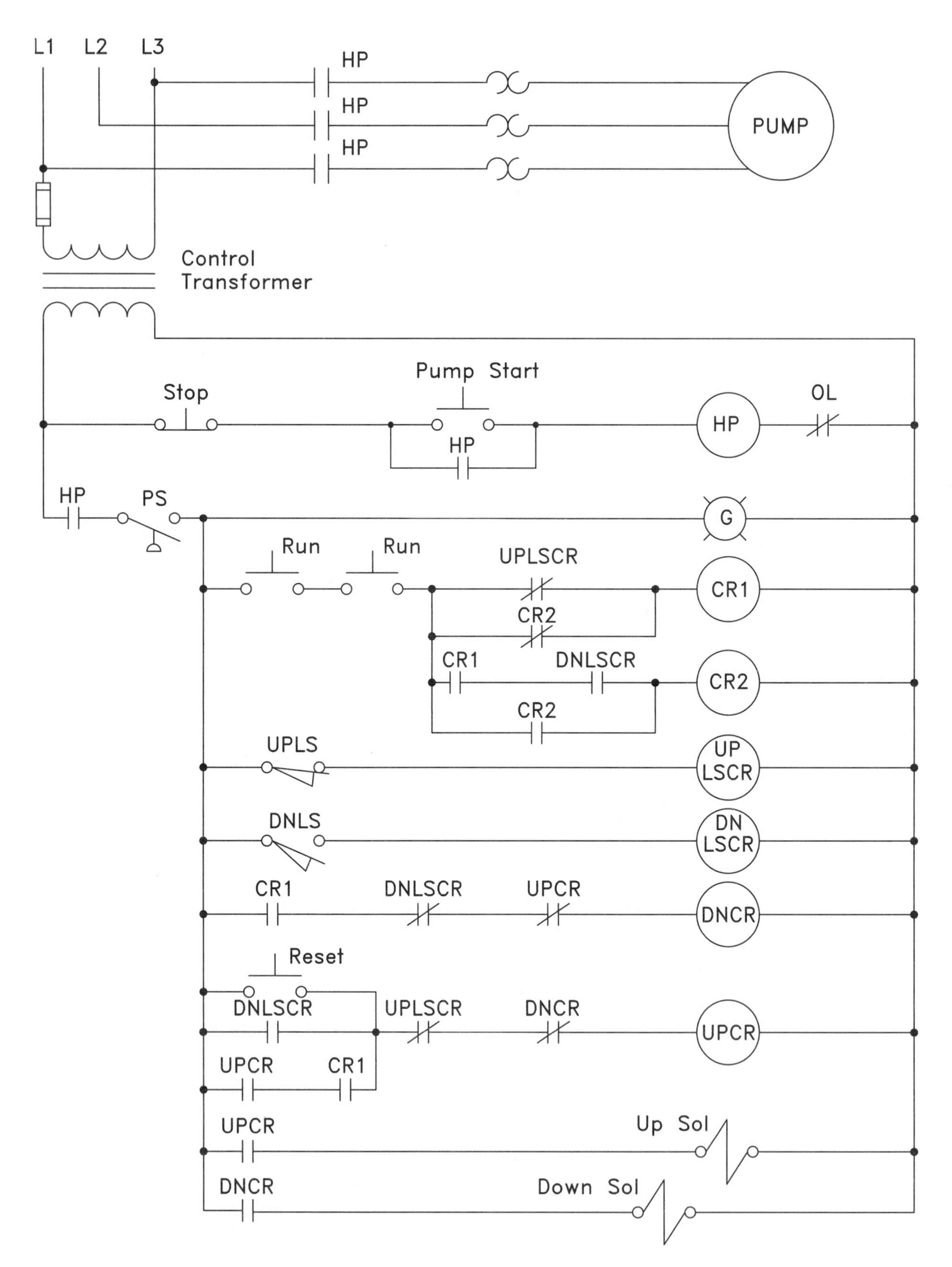

Figure 11-1 Hydraulic press

held closed. This limit switch is connected normally open, but when the press is in the up position it is being held closed. Now assume that the hydraulic pump is started and that the pressure switch closes. When pressure switch PS closes, the green pilot light turns on and UPLSCR (Up Limit Switch Control Relay) energizes, changing all UPLSCR contacts, Figure 11-2.

When both run push buttons are held down, a circuit is completed to CR1 relay, causing all CR1 contacts to change position, Figure 11-3 (page 81). The CR1 contact connected in series with the coil of DNCR closes and energizes the relay, causing all DNCR contacts to change position. The DNCR contact connected in series with the down solenoid coil closes and energizes the down solenoid.

As the press begins to move downward, limit switch UPLS opens and de-energizes coil UPLSCR, returning all UPLSCR contacts to their normal position, Figure 11-4 (page 82).

When the press reaches the bottom of its stroke, it closes down limit switch DNLS. This energizes the coil of the down limit switch control relay, DNLSCR, causing all DNLSCR contacts to change position, Figure 11-5 (page 83). The normally open DNLSCR contact connected in series with the coil of CR2 closes and energizes that relay causing all CR2 contacts to change position. The normally closed DNLSCR contact connected in series with DNCR coil opens and de-energizes that relay. All DNCR contacts return to their normal position. The normally open contact connected in series with the down solenoid coil opens and de-energizes the solenoid. The normally closed DNCR contacts connected in series with UPCR coil recloses and provides a current path to that relay.

The UPCR contact connected in series with coil DNCR opens and prevents coil DNCR from re-energizing whven coil DNLSCR de-energizes. The normally open UPCR contact connected in series with the up solenoid closes and provides a current path to the up solenoid. When the press starts upward, limit switch DNLS re-opens and de-energizes coil DNLSCR. A circuit is maintained to UPCR coil by the now closed UPCR contact connected in series with the CR1 contact, Figure 11-6 (page 84).

The press will continue to travel upward until it reaches its upper limit and closes limit switch UPLS, energizing coil UPLSCR, Figure 11-7 (page 85). This causes both UPLSCR contacts to change position. The UPLSCR contact connected in series with coil UPCR opens and de-energizes the up solenoid. Notice that control relays CR1 and CR2 are still energized. Before the press can be re-cycled, one or both of the run buttons must be released to break the circuit to the control relays. This will permit the circuit to reset to the state shown in Figure 11-2. If for some reason the press should be stopped during a cycle, the reset button can be used to return the press to the starting position.

Connecting the Circuit

In this exercise, toggle switches will be used to simulate the action of the pressure switch and the two limit switches. Lights may also be substituted for the up and down solenoid coils.

1. Refer to the circuit shown in Figure 11-1. Count the number of contacts controlled by each of the control relays to determine which should be 11 pin and which should be 8 pin. Relays that need three contacts will have to be 11 pin, and relays that need two contacts may be 8 pin.
2. After determining whether a relay is to be 11 pin or 8 pin, identify the relay with some type of marker that can be removed later. Identifying the relays as CR1, CR2, and so on, can make connection much simpler.
3. Place the pin numbers on the schematic in Figure 11-1 to correspond with the contacts and coils of the control relays. Circle the numbers to distinguish them from wire numbers.
4. Place wire numbers beside each component on the schematic.
5. Connect the circuit. (***Note:*** When connecting the two run push buttons, connect them close enough together to permit both to be held closed with one hand.)

Testing the Circuit

To test the circuit for proper operation:

1. Set the toggle switches used to simulate the pressure and down limit switch in the open (off) position. Set the toggle used to

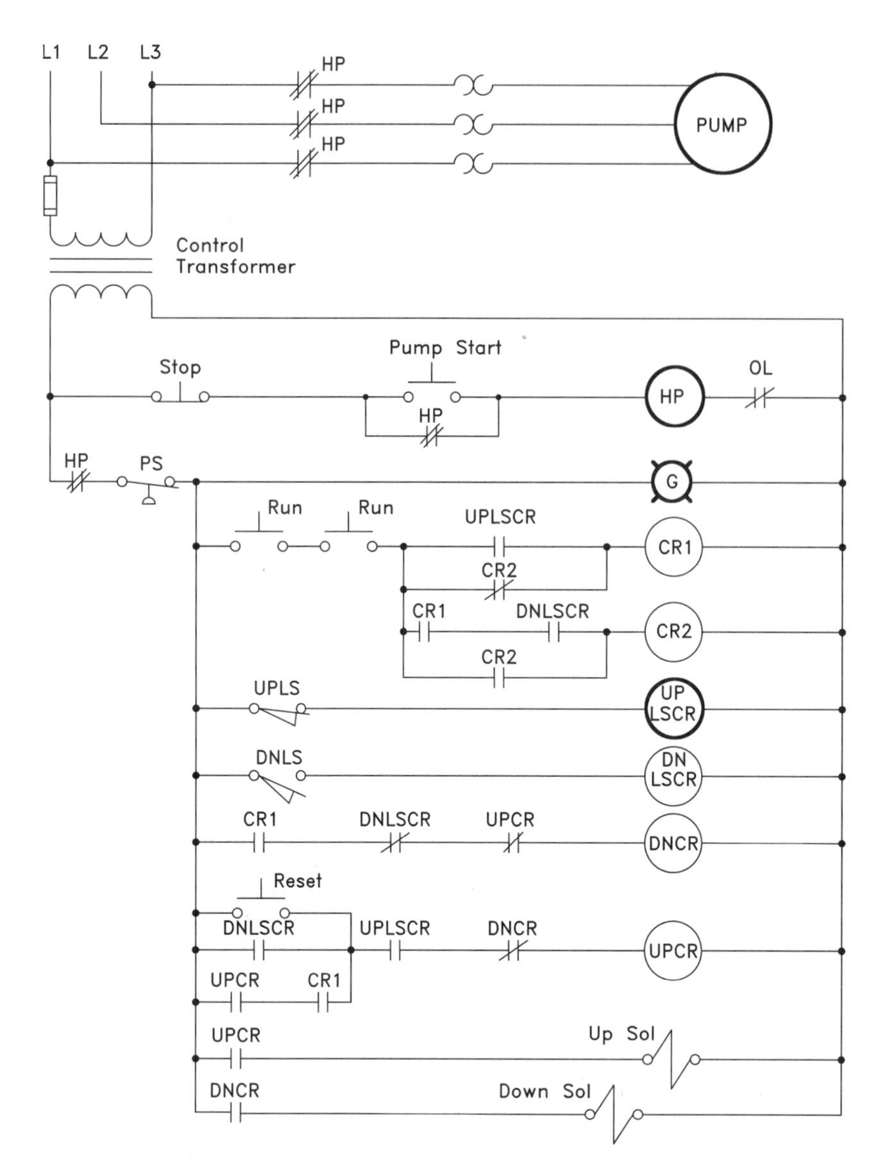

Figure 11-2 The circuit with pump operating

simulate the up limit switch in the closed (on) position.

2. Press the "pump start" button and the motor or simulated motor load should start operating.
3. Close the pressure switch. The pilot light and UPLSCR relay should energize.
4. Press and hold down both of the run push buttons. Relays CR1 and DNCR should energize. The down solenoid should also turn on.
5. The press is now traveling in the down direction. Open the up limit switch. This should cause UPLSCR to de-energize. The down solenoid should remain turned on.
6. Close the down limit switch to simulate the press reaching the bottom of its stroke. DNLSCR, CR2, and UPCR should energize. The press is now starting to travel upward.
7. Open the down limit switch. DNLSCR should de-energize, but the UPCR should remain energized.
8. Close the up limit switch to simulate the press reaching the top of its stroke. The up solenoid should turn off. Control relays CR1 and CR2 should both remain on as long as the two run buttons are held closed.
9. To restart the cycle, release the run buttons and reclose them.

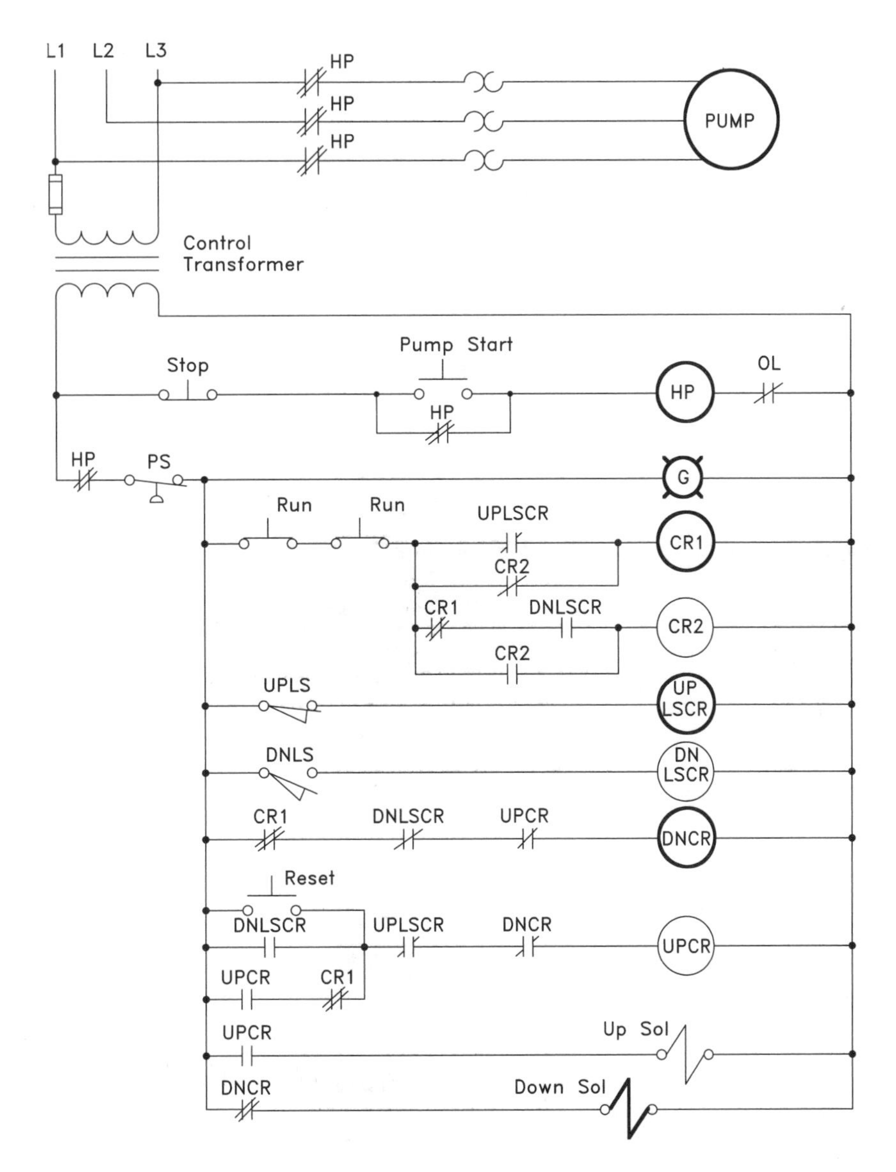

Figure 11-3 Circuit is started

Review Questions

1. Assume that the hydraulic pump is running and that the pilot light is turned on, indicating that there is sufficient pressure to operate the press. Now assume that the up limit switch is not closed. What will be the action of the circuit if both run buttons are pressed?

2. Assume that the press is in the middle of its downstroke when the operator releases the two run push buttons. Explain the action of the circuit.

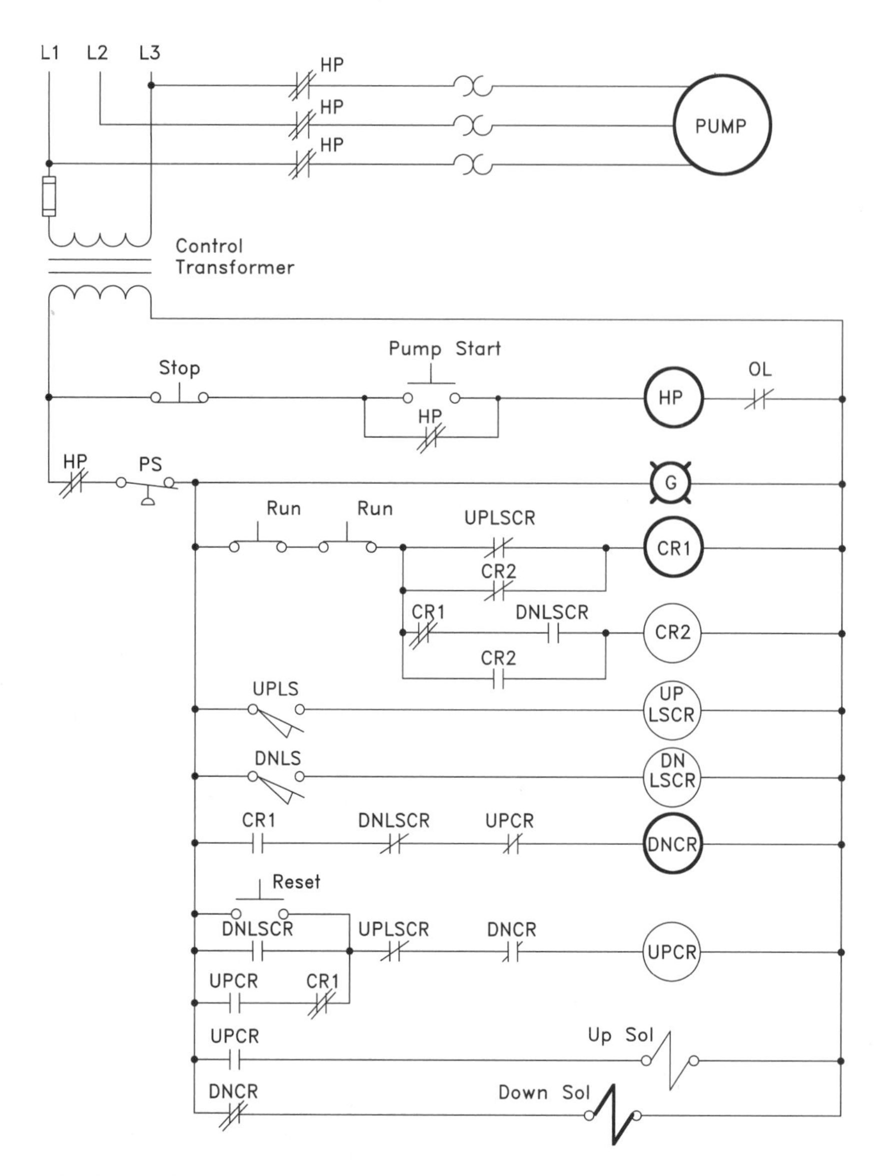

Figure 11-4 The up limit switch opens

3. Referring to the condition of the circuit as stated in review question 2, what would happen if the two run push buttons are pressed and held closed? Explain your answer.

4. Referring to the condition of the circuit as stated in review question 2, what would happen if the reset button was pressed and held closed? Explain your answer.

5. Assume that the press traveled to the bottom of its stroke and then started back up. When it reached the middle of its stroke, the power was interrupted. After the power has been restored,

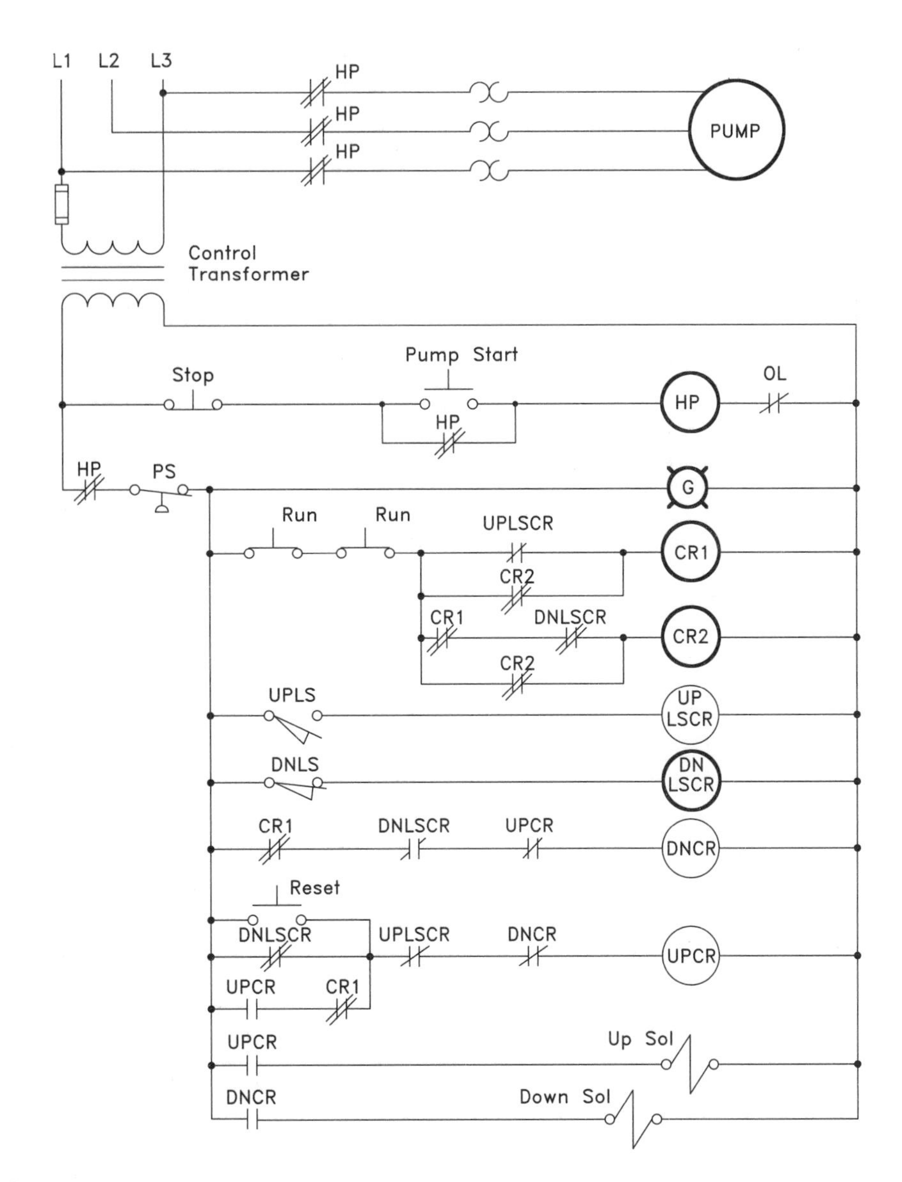

Figure 11-5 DNLSCR and CR2 relays energize

if the two run buttons are pressed, will the press continue to travel upward to complete its stroke, or will it start moving downward?

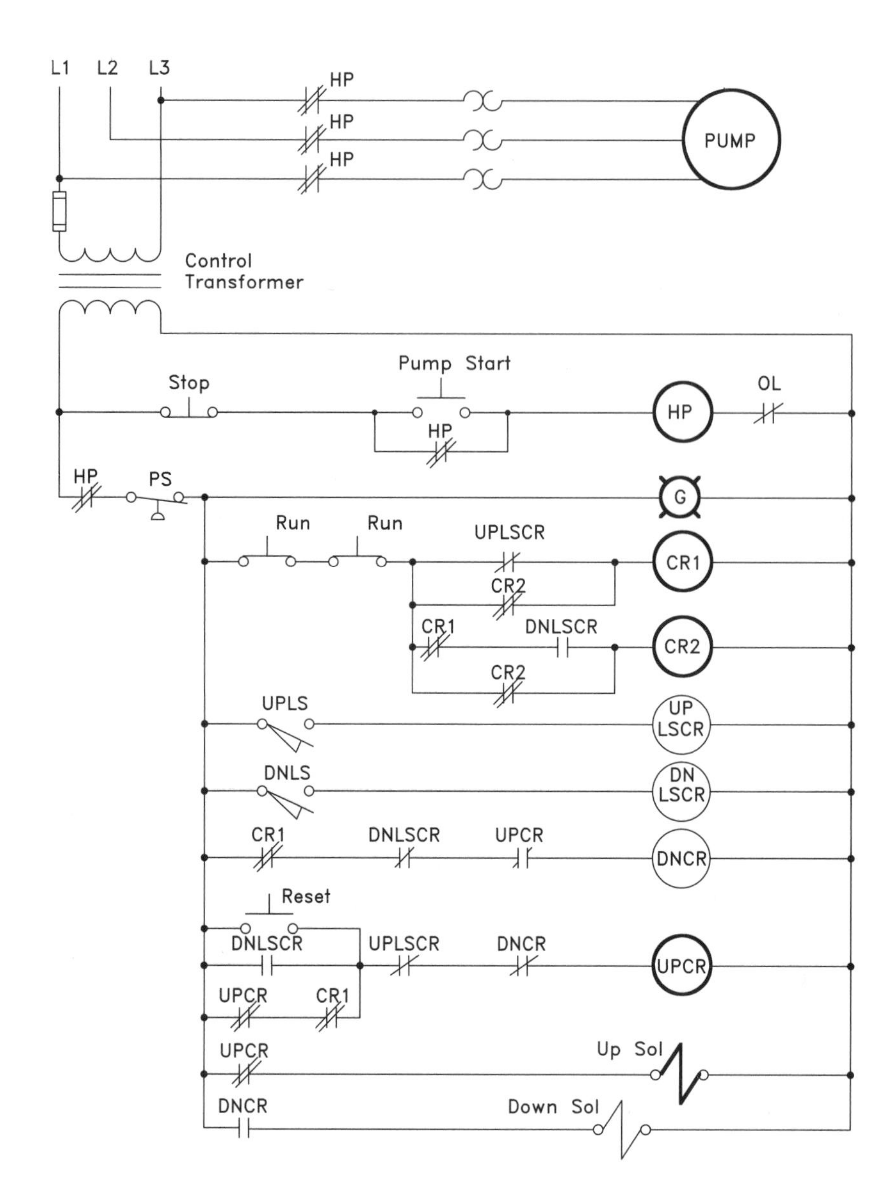

Figure 11-6 Limit switch DNLS re-opens

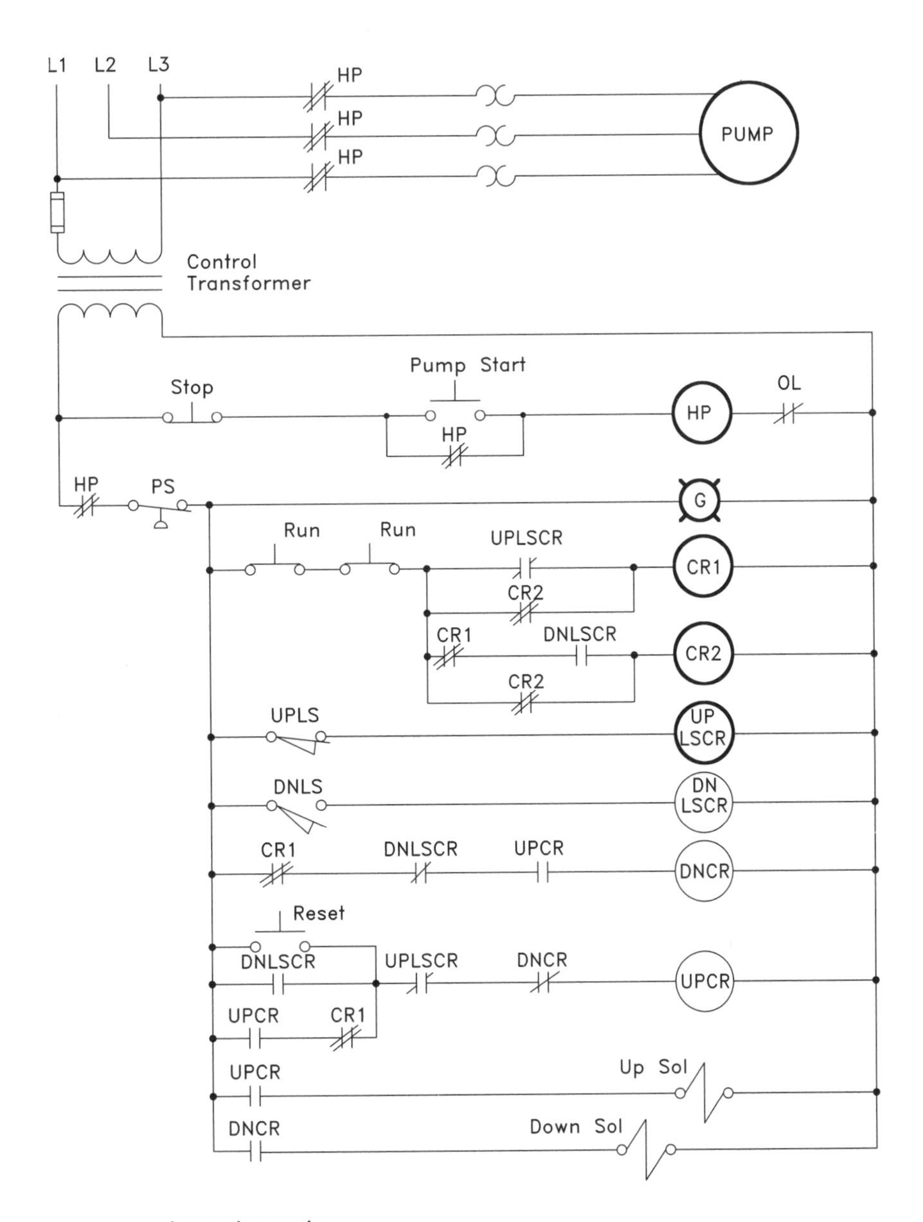

Figure 11-7 The press completes the cycle

EXPERIMENT 12

Design of Two Flashing Lights

Name: Date: Grade:

Comments:

Objectives

After completing this experiment, the student should be able to:

- Design a circuit from a written statement of requirements
- Connect the circuit in the laboratory after the design has been approved

Materials Needed

Materials depend on the circuit design.

In the space provided in Figure 12-1, draw a schematic diagram of a circuit that will fulfill the following requirements. Use two separate timers. Do not use an electronic timer set in the repeat mode. Remember that there is generally more than one way to design any circuit. Try to keep the design as simple as possible. The fewer components a circuit has, the less likely it is to fail.

1. An on-off toggle switch is used to connect power to the circuit.
2. When the switch is turned on, two lights will alternately flash on and off. Light 1 will be turned on when light 2 is turned off. When light 1 turns off, light 2 will turn on.
3. The lights are to flash at a rate of on for one second and off for one second.

When completed, have your instructor approve the design. After the design has been approved, connect the circuit in the laboratory.

Review Questions

1. When designing a control circuit that requires the use of a timing relay, what type of timer is generally used during the design?

2. Should schematic diagrams be drawn to assume that the circuit is energized or de-energized?

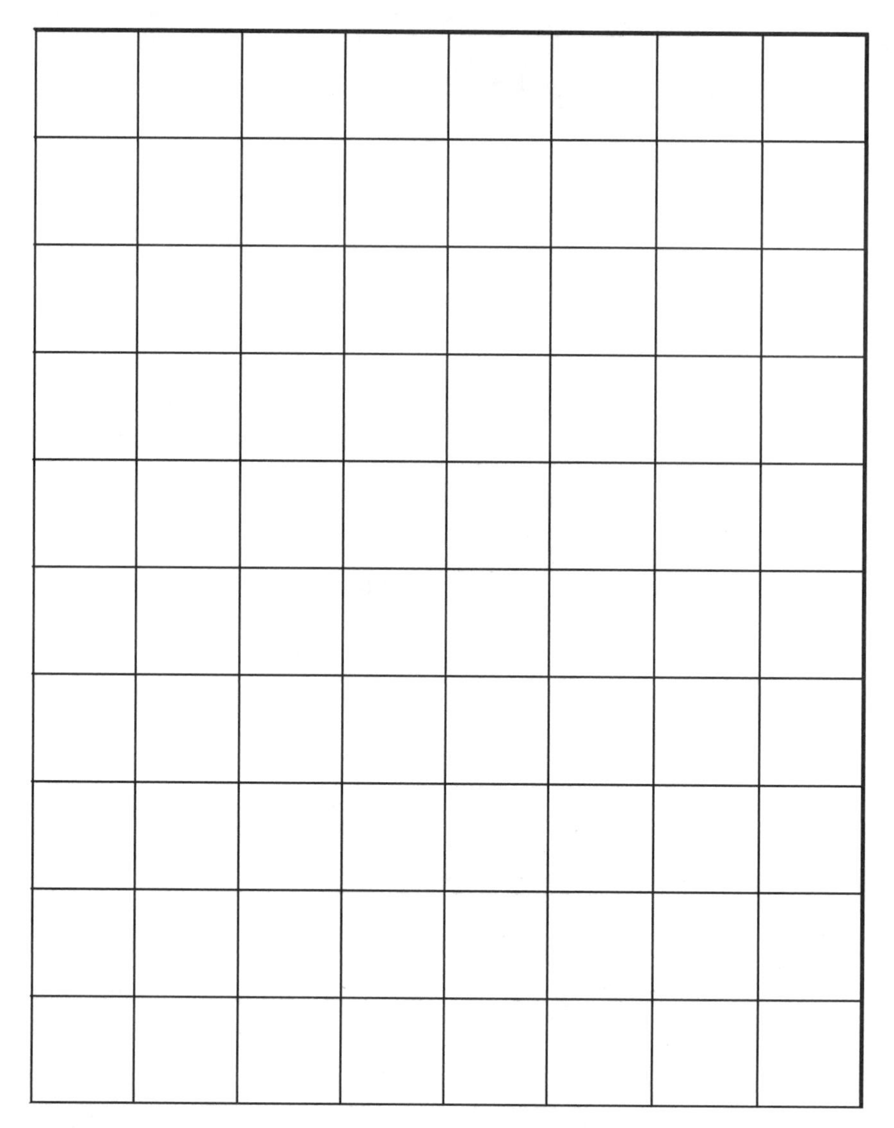

Figure 12-1 Design of two flashing lights

3. Explain the difference between a schematic and a wiring diagram.

4. In a forward-reverse control circuit, a normally closed F contact is connected in series with the R starter coil, and a normally closed R contact is connected in series with the F starter coil. What is the purpose of doing this and what is the contact arrangement called?

5. What type of overload relay is not sensitive to changes in ambient temperature?

EXPERIMENT 13

Design of Three Flashing Lights

Name:	Date:	Grade:
Comments:		

Objectives

After completing this experiment, the student should be able to:

- Design a motor control circuit using timers
- Discuss the operation of this circuit
- Connect this circuit in the laboratory

Materials Needed

Materials depend on the design of the circuit.

The design of this circuit will be somewhat similar to the circuit in Experiment 12. This circuit, however, contains three lights that turn on and off in sequence. Use the space provided in Figure 13-1 to design this circuit. The requirements of the circuit are as follows:

1. A toggle switch is used to connect power to the circuit. When the power is turned on, light 1 will turn on.
2. After a delay of one second, light 1 will turn off and light 2 will turn on.
3. After a delay of one second, light 2 will turn off and light 3 will turn on.
4. After a delay of one second, light 3 will turn off and light 1 will turn back on.
5. The lights will repeat this action until the toggle switch is opened.

Procedure

1. After the design of your circuit has been approved by your instructor, connect the circuit in the laboratory.
2. Test the circuit for proper operation.
3. Disconnect the circuit and return the components to their proper location.

Review Questions

1. A 60 hp three-phase squirrel cage induction motor is to be connected to a 480 volt line. What size NEMA starter should be used to make this connection?

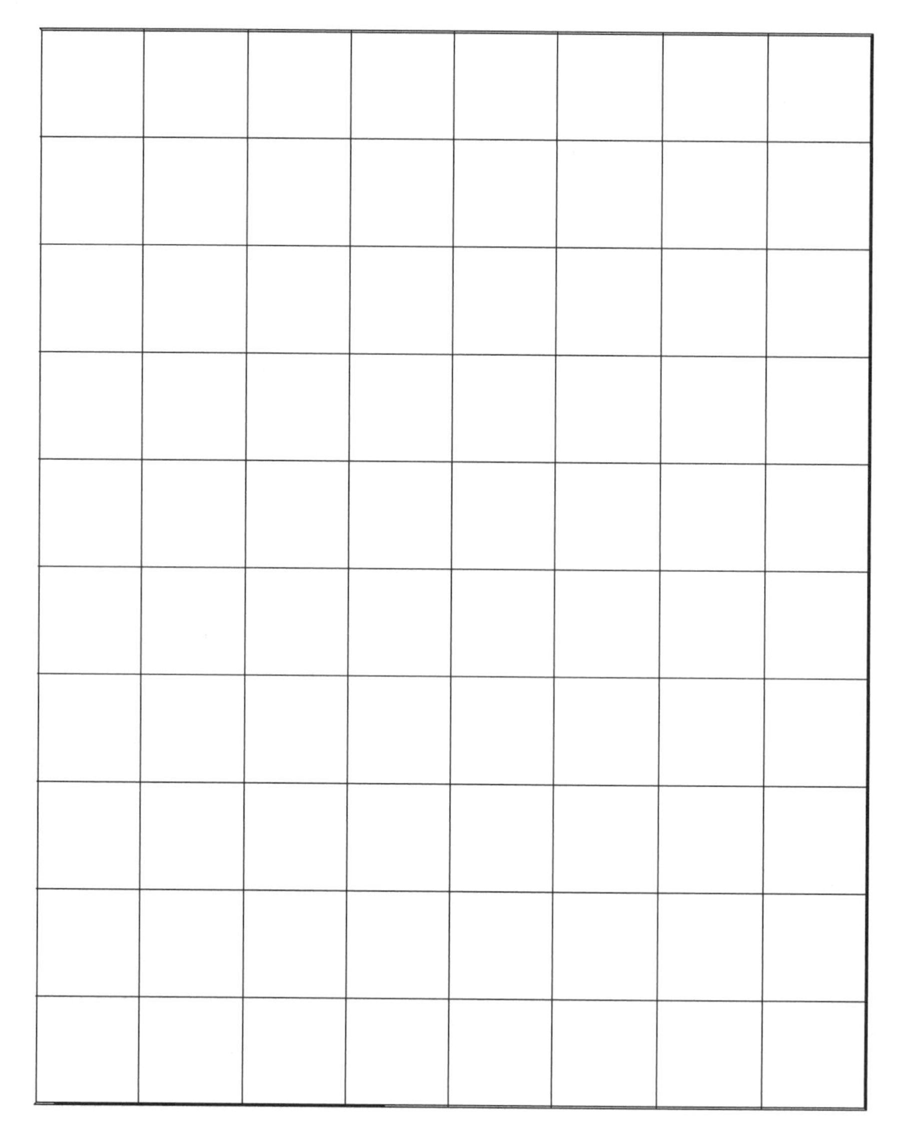

Figure 13-1 Design of three lights that turn on and off in sequence

2. An electrician is given a NEMA size 2 starter to connect a 30 hp three-phase squirrel cage motor to a 575 volt line. Should this starter be used to operate this motor?

3. Assume that the motor in review question 2 has a design code B. What standard size inverse time circuit breaker should be used to connect the motor?

4. The motor described in review questions 2 and 3 is to be connected with copper conductors with type THHN insulation. What size conductors should be used? The termination temperature rating is not known.

5. Assume that the motor in review question 2 has a nameplate current rating of 28 amperes, and a marked service factor of 1. What size overload heater should be used for this motor?

EXPERIMENT 14

Control for Three Pumps

Name: Date: Grade:

Comments:

Objectives

After completing this experiment, the student should be able to:

- Analyze a motor control circuit
- List the steps of operation in a control circuit
- Connect this circuit in the laboratory

Materials Needed

- Three-phase power supply
- Control transformer
- Three motor starters with normally open auxiliary contacts
- 6 ea.—Toggle switches to simulate auto-man switches and float switches
- 8 pin control relay and 8 pin socket
- 3 ea.—Three-phase motors or equivalent motor loads
- One normally open and one normally closed push button

One of the primary duties of an industrial electrician is to troubleshoot existing control circuits. To troubleshoot a circuit, the electrician must understand what the circuit is designed to do and how it accomplishes it. To analyze a control circuit, start by listing the major components. Next, determine the basic function of each component. Finally, determine what occurs during the circuit operation.

To illustrate this procedure, the circuit previously discussed in Experiment 11 will be analyzed. The hydraulic press circuit is shown in Figure 14-1. In order to facilitate circuit analysis, wire numbers have been placed beside the components. The first step will be to list the major components in the circuit:

1. Normally closed stop push button
2. Normally open push button used to start the hydraulic pump
3. 2 ea.—Normally open push buttons used as run buttons
4. Normally open push button used for the reset button
5. Normally open pressure switch
6. 2 ea.—Normally open limit switches
7. 2 ea.—Solenoid valves
8. 3 ea.—8 pin control relays (CR2, UPLSCR, and DNCR)
9. 3 ea.—11 pin control relays (CR1, DNLSCR, and UPCR)
10. Control transformer
11. Green pilot light

The next step in the process is to give a brief description of the function of each listed component:

1. (Normally closed stop push button)—Used to stop the operation of the hydraulic pump motor.
2. (Normally open push button used to start the hydraulic pump)—Starts the hydraulic pump.

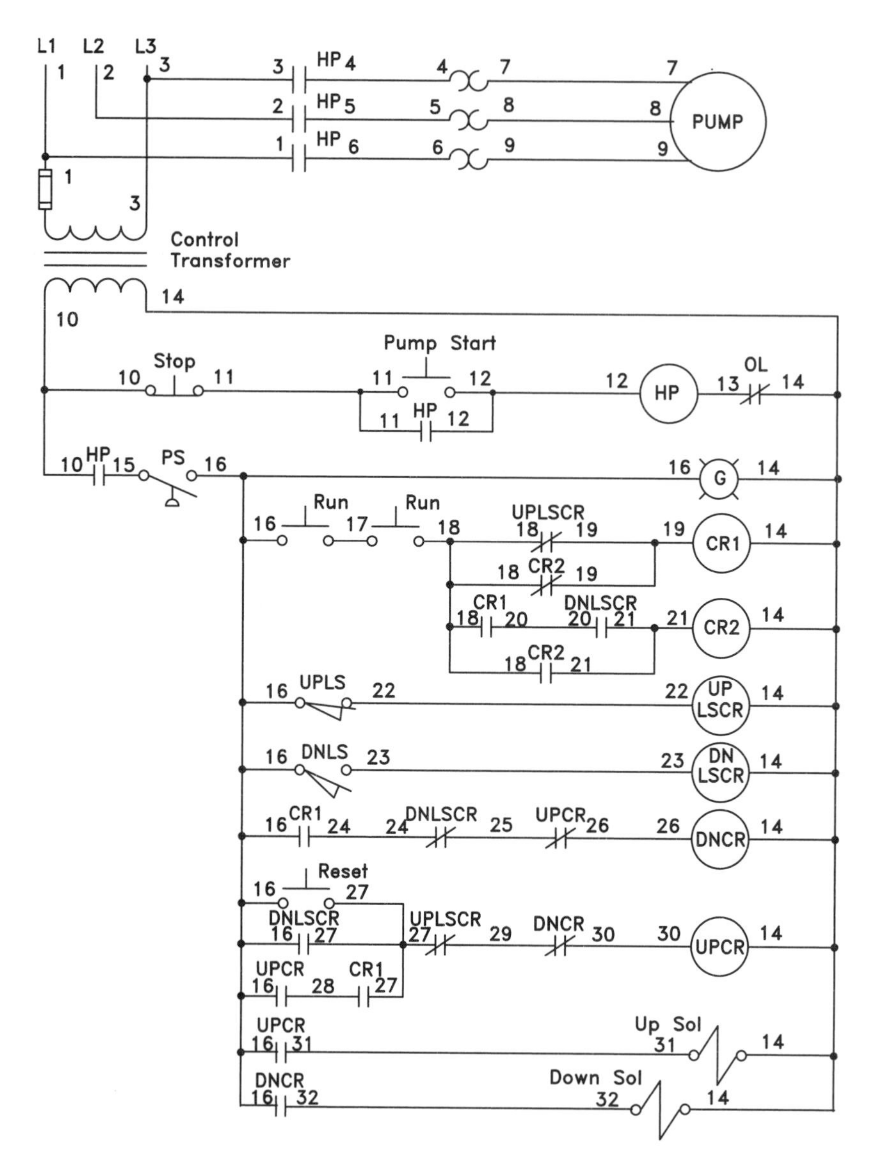

Figure 14-1 Analyzing the circuit

3. (2 ea.—Normally open push buttons used as run buttons)—Both push buttons must be held down to start the action of the press.
4. (Normally open push button used for the reset button)—Resets the press to the topmost position.
5. (Normally open pressure switch)—Determines whether or not there is enough hydraulic pressure to operate the press.
6. (2 ea.—Normally open limit switches)—Determine when the press is at the top of its stroke and when it is at the bottom of its stroke.
7. (2 ea.—Solenoid valves)—The up solenoid valve opens on energize to permit hydraulic fluid to move the press upward. The down solenoid valve opens on energize to permit hydraulic fluid to move the press downward.
8. (3 ea.—8 pin control relays [CR2, UPLSCR, and DNCR])—Part of the control circuit.
9. (3 ea.—11 pin control relays [CR1, DNLSCR, and UPCR])—Part of the control circuit.
10. (Control transformer)—Reduces the value of the line voltage to the voltage needed to operate the control circuit.
11. (Green pilot light)—Indicates that there is enough hydraulic pressure to operate the pump.

The final step is to analyze the operation of the circuit. To analyze circuit operation, trace the current paths each time a change is made in the circuit. Start by pressing the pump start button.

1. When the pump start button is pressed, a circuit is completed to the coil of starter HP.
2. When coil HP energizes, all HP contacts change position. The three load contacts close to connect the pump motor to the line. The HP auxiliary contact located between wire points 11 and 12 closes to maintain the circuit after the pump start button is released, and the HP auxiliary contact located between wire numbers 10 and 15 closes to provide power to the rest of the circuit.
3. After the hydraulic pump starts, the hydraulic pressure in the system increases and closes the pressure switch.
4. When the pressure switch closes, a current path is provided to the green pilot light to indicate that there is sufficient hydraulic pressure to operate the press. A current path also exists through the normally open held closed up limit switch to control relay coil UPLSCR.
5. When UPLSCR relay energizes, both UPLSCR contacts open. The UPLSCR contact located between wire numbers 18 and 19 opens to break a current path to CR1 coil. UPLSCR contact located between wire numbers 27 and 29 opens to break the current path to coil UPCR.
6. Both run push buttons must be held down to provide a current path through the normally closed CR2 contact located between wire numbers 18 and 19 to the coil of CR1 relay.
7. When CR1 relay coil energizes, the CR1 contact located between wire numbers 18 and 20 closes to provide a path to CR2 coil in the event that the DNLSCR contact should close. The CR1 contact located between wire numbers 16 and 24 closes to provide a current path to the down control relay (DNCR). The CR1 contact located between wire numbers 28 and 27 closes to provide an eventual current path to the up control relay (UPCR).
8. When DNCR coil energizes, the DNCR contact located between wire numbers 29 and 30 opens to provide interlock with the up control relay. The DNCR contact between wire numbers 16 and 32 closes and provides a current path to the down solenoid valve.
9. When the down solenoid valve energizes, the press begins its downward stroke. This causes the normally open held closed up limit switch to open and de-energize the coil of the up limit switch control relay (UPLSCR).
10. Both UPLSCR contacts reclose.
11. When the press reaches the bottom of its stroke, the down limit switch located between wire numbers 16 and 23 closes to provide a current path to the coil of the down limit switch control relay (DNLSCR).
12. All DNLSCR contacts change position. The DNLSCR contacts located between wire numbers 20 and 21 close to provide a current path though the now closed CR1 contact to the coil of CR2 relay. The DNLSCR contact located between wire numbers 24 and 25 opens and breaks the current path to DNCR relay. The DNLSCR contact located between wire numbers 16 and 27 closes to provide a current path to UPCR relay when the DNCR contact located between 29 and 30 recloses.
13. When CR2 coil energizes, the normally closed CR2 contact located between wires 18 and 19 opens to prevent a maintained current path to CR1 when the UPLSCR contact reopens. The normally open CR2 contact located between 18 and 21 closes to maintain a current path to the coil of CR2 in the event that CR1 or DNLSCR contacts should open.
14. When the DNCR relay coil de-energizes, the DNCR contact located between wires 29 and 30 recloses to permit coil UPCR to be energized. The DNCR contact located between 16 and 32 reopens to break the current path to the down solenoid valve.
15. When the UPCR coil energizes, the normally closed UPCR contact located between wires 25 and 26 opens to provide interlock with the DNCR relay coil. The UPCR contact located between 16 and 28 closes to maintain a circuit through the now closed CR1 contact to the coil of UPCR. The UPCR contact located between 16 and 31 closes and provides a current path to the up solenoid valve.

16. When the up solenoid valve opens, hydraulic fluid causes the press to begin its upward stroke.
17. When the press starts upward, the down limit switch reopens and de-energizes the coil of DNLSCR relay.
18. When coil DNLSCR de-energizes, the DNLSCR contact located between wires 20 and 21 reopens, but a current path is maintained by the now closed CR2 contact. The DNLSCR contact located between 24 and 25 recloses, but the current path to DNCR coil remains broken by the UPCR contact located between 25 and 26. The DNLSCR contact located between wires 16 and 27 reopens, but a current path is maintained by the now closed UPCR and CR1 contacts.
19. When the press reaches the top of its stroke, the up limit switch again closes and provides a current path to the coil of UPLSCR relay.
20. The UPLSCR contact located between wires 18 and 19 opens to break the current path to CR1 coil. The UPLSCR contact located between wires 27 and 29 opens to break the current path to the coil of UPCR.
21. When CR1 coil de-energizes, all CR1 contacts return to their normal position. The CR1 contact between wires 18 and 20 reopens, CR1 contact between wires 16 and 24 reopens to prevent a current path from being established to the DNCR relay coil, and CR1 contact between wires 27 and 28 reopens.
22. When coil UPCR de-energizes, its contacts return to their normal position. The UPCR contacts located between wires 16 and 28 reopen, and the UPCR contact located between wires 16 and 31 reopens to break the circuit to the up solenoid.
23. Before the circuit can be restarted, the current path to relay CR2 must be broken by releasing one or both of the run push buttons. This will return all contacts back to their original state.
24. In the event the press should be stopped in the middle of its stroke, the up limit switch will be open and coil UPLSCR will be de-energized. The DNCR coil will also be de-energized. If the reset button is pressed and held, a circuit will be completed through the normally closed DNLSCR and DNCR contacts to the coil of UPCR. This will cause the up solenoid valve to energize and return the press to its up position.

Determining What the Circuit Does

The circuit in this experiment is intended to operate three pumps. The pumps are used to pump water from a sump to a roof storage tank. The water in the storage tank is used for cooling throughout the plant. After the water has been used for cooling, it returns to the sump to be recooled. Three float switches are used to detect the water level in the storage tank. As the water is drained out of the tank, the level drops and the float switches turn on the pumps to pump water from the sump back to the storage tank, Figure 14-2.

List the Components

In the space provided, list the major components in the control circuit shown in Figure 14-3:

1.

2.

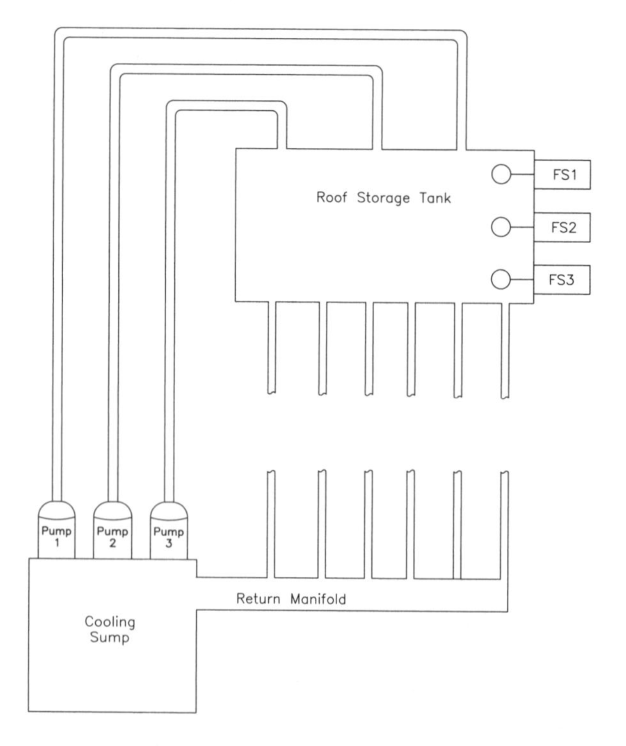

Figure 14-2 Roof mounted tank for plant cooling system

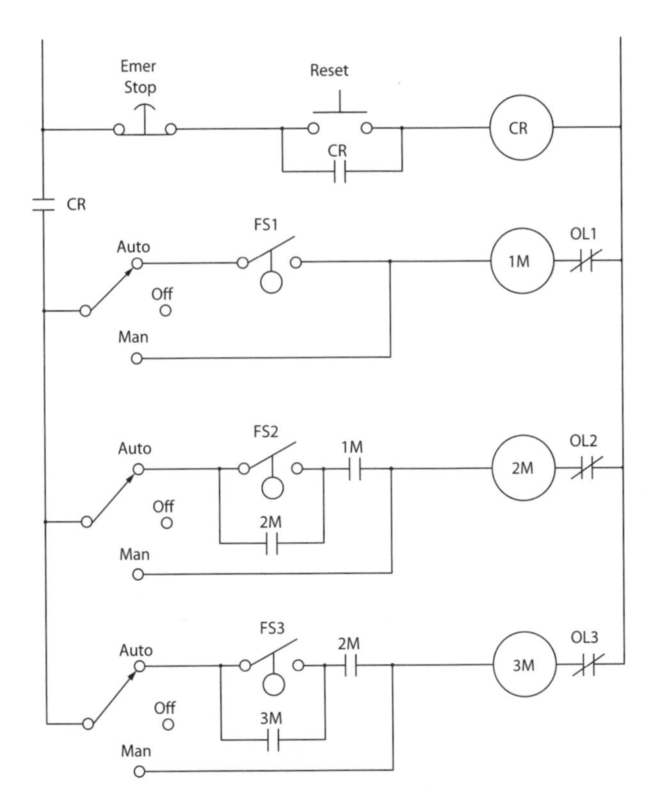

Figure 14-3 Control circuit for three pumps

3.

4.

5.

6.

7.

8.

9.

10.

Describe the Components

In the space provided, give a brief description of the function of the components in this circuit:

1.

2.

3.

4.

5.

6.

7.

8.

9.

10.

Describing the Circuit Operation

In the space provided, describe the operation of the circuit. Assume that in the normal state the roof storage tank is filled with water, and all the auto-off-man switches are set in the auto position. Also assume that the three motor starters control the operation of the three pumps, although the pumps are not shown on the schematic.

1.

2.

3.

4.

5.

6.

7.

8.

9.

10.

11.

12.

13.

14.

15.

Review Questions

To answer the following questions, refer to the circuit shown in Figure 14-3.

1. Assume that all three pumps are operating. What would be the action of the circuit if the auto-off-man switch of pump 2 were to be switched to the off position?

2. Assume that the auto-off-man switch of pump 3 is set in the manual position. What will be the operation of the circuit if float switch FS1 closes?

3. Assume that the roof storage tank empties completely, but none of the pumps has started. Which of the following could **not** cause this condition?
   a. The emergency stop button has been pushed, and the control relay is de-energized.
   b. The auto-off-man switch of pump 1 has been set in the off position.
   c. The auto-off-man switch of pump 1 has been set in the manual position.
   d. 1M coil is open.
4. Assume that all three pumps are in operation and OL3 contact opens. Will this affect the operation of the other two pumps?

5. Assume that FS2 float switch is defective. If the water level drops enough to close float switch FS3, will pump 3 start running?

EXPERIMENT 15

Oil Pressure Pump Circuit for a Compressor

Name: Date: Grade:

Comments:

Objectives

After completing this experiment the student should be able to:

- Analyze a motor control circuit
- List the steps of operation in a control circuit
- Connect this circuit in the laboratory

Materials Needed

- Three-phase power supply
- 2 ea.—Motor starters
- Control transformer
- 2 ea.—Electronic timers (Dayton model 6A855) and 11 pin tube sockets
- 2 ea.—Pilot lights
- 2 ea.—Double acting push buttons

In the circuit shown in Figure 15-1, the oil pump must start for some time before the compressor is started. When the start button is pressed, the oil pump should continue to run for some time after the compressor stops operating.

Listing the Components

In the space provided, list the circuit components:

1.
2.
3.
4.
5.
6.
7.
8.
9.
10.

Describe the Components

In the space provided, give a brief description of what function is performed by each component:

1.

2.

3.

4.

5.

6.

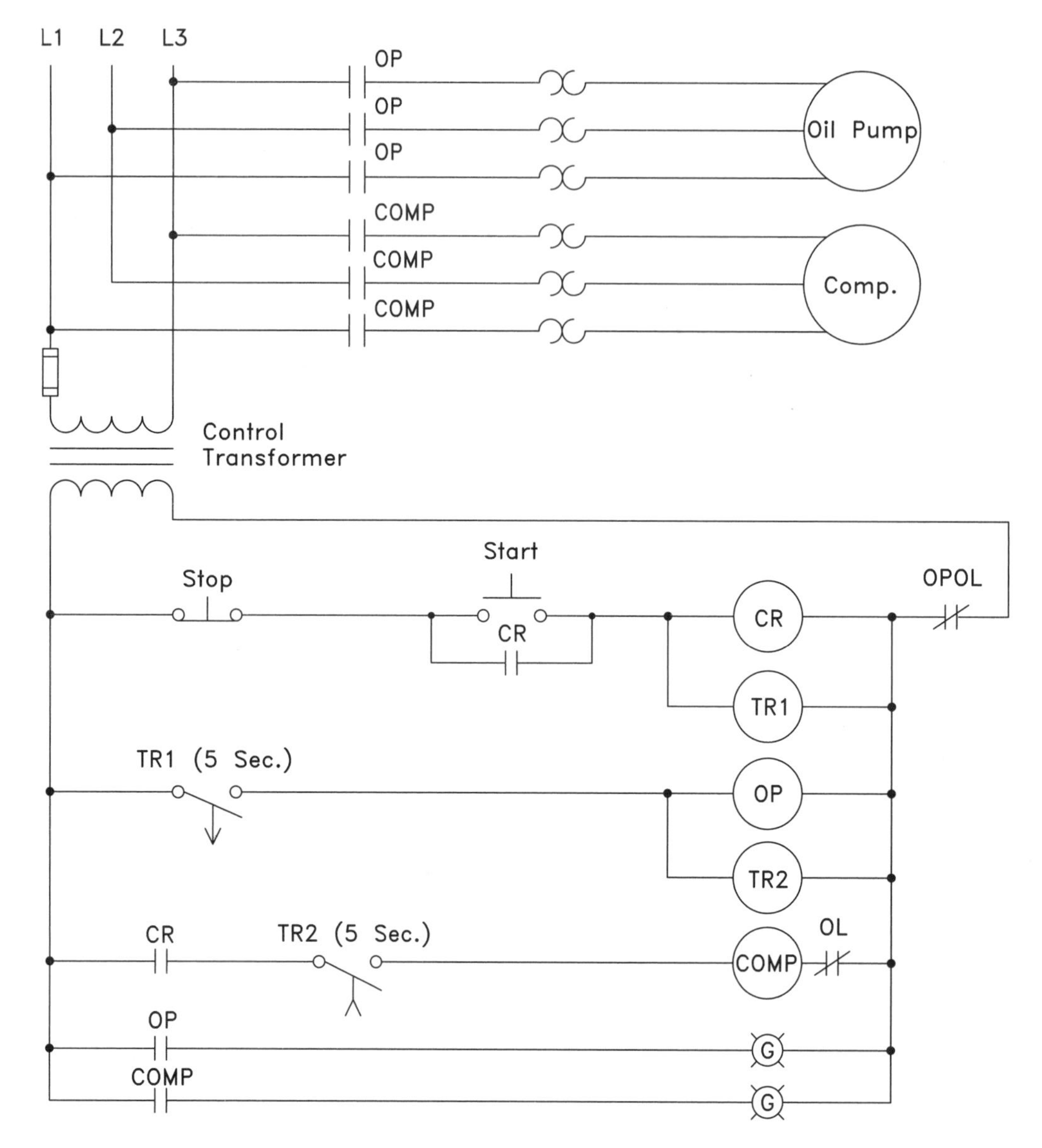

Figure 15-1 Compressor oil pump circuit

7.

8.

9.

10.

Circuit Operation

In the space provided, describe the operation of the circuit in a step-by-step sequence:

1.

2.

3.

4.

5.

6.

7.

8.

9.

10.

11.

12.

13.

14.

15.

Connecting the Circuit

1. Connect the circuit shown in Figure 15-1.
2. After checking with the instructor, turn on the power and test the circuit for proper operation.
3. Turn off the power and disconnect the circuit. Return the components to their proper location.

Review Questions

To answer the following questions, refer to the circuit shown in Figure 15-1.

1. Assume that the start button is pressed and the oil pump starts operating. After a delay of five seconds, the COMP pilot light turns on, but the compressor motor does not start. Which of the following could cause this condition?
   a. TR2 timer is defective.
   b. COMP starter coil is defective.
   c. The compressor motor is defective.
   d. All the above.

2. Assume that the circuit is in operation. When the stop button is pressed, both the compressor and the oil pump stop operating immediately. Which of the following could cause this condition?
   a. CR relay is defective.
   b. TR1 timer is defective.
   c. OP starter is defective.
   d. Timer TR2 is defective.

3. When the start button is pressed, the oil pump starts operating immediately. After a delay of five seconds, the oil pump motor turns off. An electrician finds that the control transformer fuse is blown. Which of the following could cause this condition?
   a. TR1 coil is shorted.
   b. OP coil is shorted.
   c. TR2 coil is shorted.
   d. COMP coil is shorted.

4. When the start button is pressed, the oil pump motor starts operating immediately. After a long time delay, it is determined that the compressor motor will not start. Which of the following could **not** cause this condition?
   a. OP coil is defective.
   b. TR2 coil is defective.
   c. COMP coil is defective.
   d. The compressor overload contact is open.

5. When the start button is pressed, the oil pump motor starts operating immediately. When the start button is released, however, the oil pump motor turns off. The operator then presses the start button and holds it down for a period of ten seconds. This time the oil pump motor starts operating immediately, but the compressor motor never starts. When the start button is released, the oil pump motor again immediately turns off. Which of the following could cause this condition?
   a. CR coil is defective.
   b. TR1 coil is defective.
   c. TR2 coil is defective.
   d. COMP coil is defective.

EXPERIMENT 16

Auto-Transformer Starter

Name:	Date:	Grade:
Comments:		

Objectives

After completing this experiment, the student should be able to:

- Discuss the operation of an auto-transformer starter
- Explain the operation of an auto-transformer starter
- Connect an auto-transformer starter in the laboratory

Materials Needed

- Three-phase power supply
- Control transformer
- 3 ea.—Three-phase contactors with at least one normally open and one normally closed auxiliary contact
- 2 ea.—0.5 KVA control transformer (480/240 - 120)
- Three-phase motor or equivalent motor load
- On-delay timer—Dayton model 6A855 or equivalent and 11 pin tube socket
- 8 pin control relay and 8 pin tube socket
- 2 ea.—Double acting push buttons (N.O./N.C. on each button)
- Three-phase overload relay or three single-phase overload relays with the overload contacts connected in series.

Auto-transformer starters are used to reduce the amount of inrush current when starting a large motor. The auto-transformer starter accomplishes this by reducing the voltage applied to the motor during the starting period. If the voltage is reduced by half, the current will be reduced by half, and the torque will be reduced to one quarter of normal.

There are several different ways to construct an auto-transformer starter. Some use three transformers, and others use two transformers. In this experiment, two transformers connected as an open delta will be used. Two 0.5 KVA control transformers will be employed. Because these transformers are to be used as auto-transformers, only the high-voltage windings will be connected. The low-voltage windings (X1 and X2) will not be used in this experiment. The high-voltage windings can be identified by the markings on the terminal leads of H1 through H4. These high-voltage windings are to be connected in series by connecting a jumper between terminals H2 and H3. This jumpered point provides a center tap for the entire winding.

Obtaining Enough Contacts

A schematic diagram of this connection is shown in Figure 16-1. Notice that a total of five S contacts are needed during the starting period. Contactors that contain five load contacts can be purchased, but they are difficult to obtain and

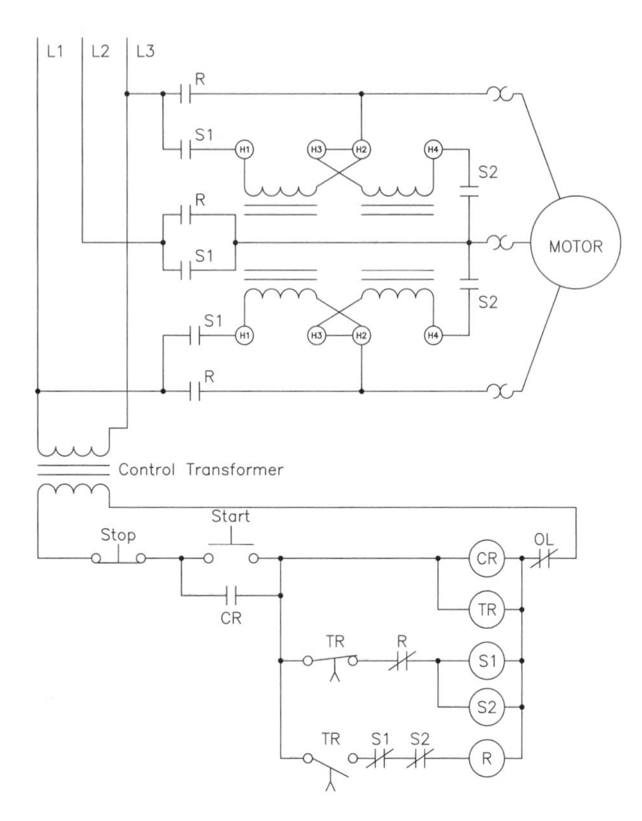

Figure 16-1 Auto-transformer starter

they are expensive. For this reason, two three-phase contactors will be employed to provide the needed load contacts. This can be accomplished by connecting the coils of S1 and S2 contactors in parallel with each other.

Circuit Operation

When the start button is pressed, coils CR, TR, S1, and S2 energize. When the S1 and S2 load contacts close, the motor is connected to the center tap of the open delta auto-transformer. Because the transformers have been center tapped, the motor is connected to one half of the line voltage. A basic schematic diagram of this connection is shown in Figure 16-2. The normally closed S1 and S2 auxiliary contacts connected in series with R coil open to provide interlock and prevent R contactor from energizing as long as S1 or S2 is energized.

After some period of time, TR timer reaches the end of its timing sequence and the two timed TR contacts change position. The normally closed TR contact connected in series with coils S1 and S2 opens and de-energizes these contactors. This causes all S1 and S2 load contacts to open and disconnect the auto-transformer from the line. The normally closed S1 and S2 auxiliary contacts connected in series with R coil reclose.

When the normally open TR contact connected in series with R coil closes, R contactor energizes and closes all R load contacts. This connects the motor directly to the power line. The normally closed R auxiliary contact connected in series with coils S1 and S2 opens to provide interlock. The motor will continue to run until the stop button is pressed or an overload occurs.

Connecting the Circuit

1. Assuming that relay CR is an 8 pin control relay and that timer TR is a Dayton model 6A855, place pin numbers beside the components of CR and TR in Figure 16-1. Circle the pin numbers to distinguish them from wire numbers.
2. Place wire numbers beside all circuit components in Figure 16-1.
3. Place corresponding wire numbers beside the components shown in Figure 16-3. Make certain to make the connection between H2 and H3 on the high-voltage side of the control transformers.
4. Connect the control section of the circuit shown in Figure 16-1.
5. Set the timing relay for a delay of five seconds.
6. After checking with the instructor, turn on the power and test the control section of the circuit for proper operation.
7. Turn off the power.
8. Connect the load section of the circuit.
9. Turn on the power and test the circuit for proper operation. (***Note:*** Connect a voltmeter across the motor or equivalent motor load terminals and monitor the voltage. When the circuit is first energized, the voltage applied to the motor should be one-half the full line value. After a delay of five seconds, the voltage should increase to full value.)
10. Turn off the power and disconnect the circuit. Return the components to their proper place.

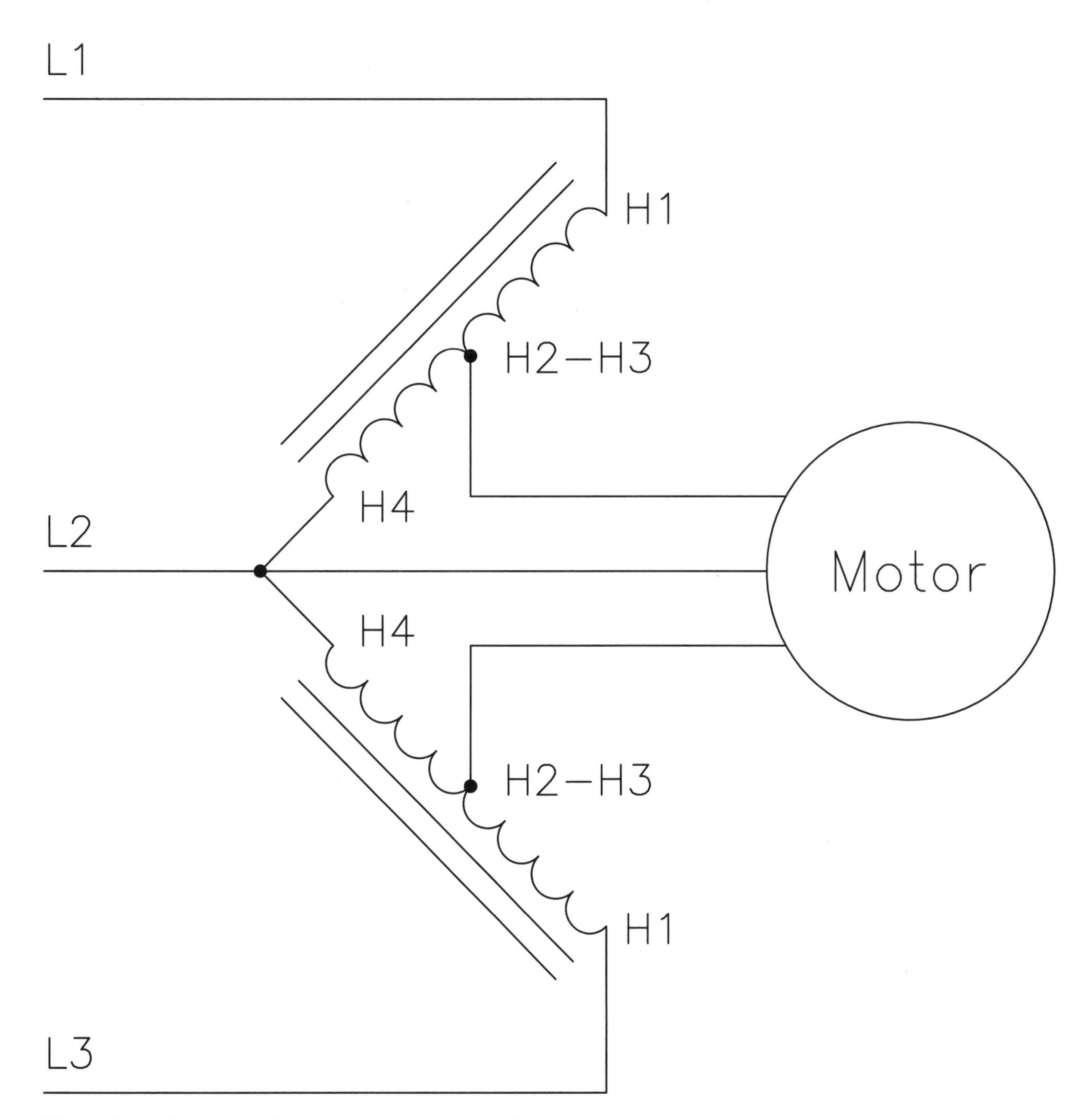

Figure 16-2 Schematic diagram of basic auto-transformer connection

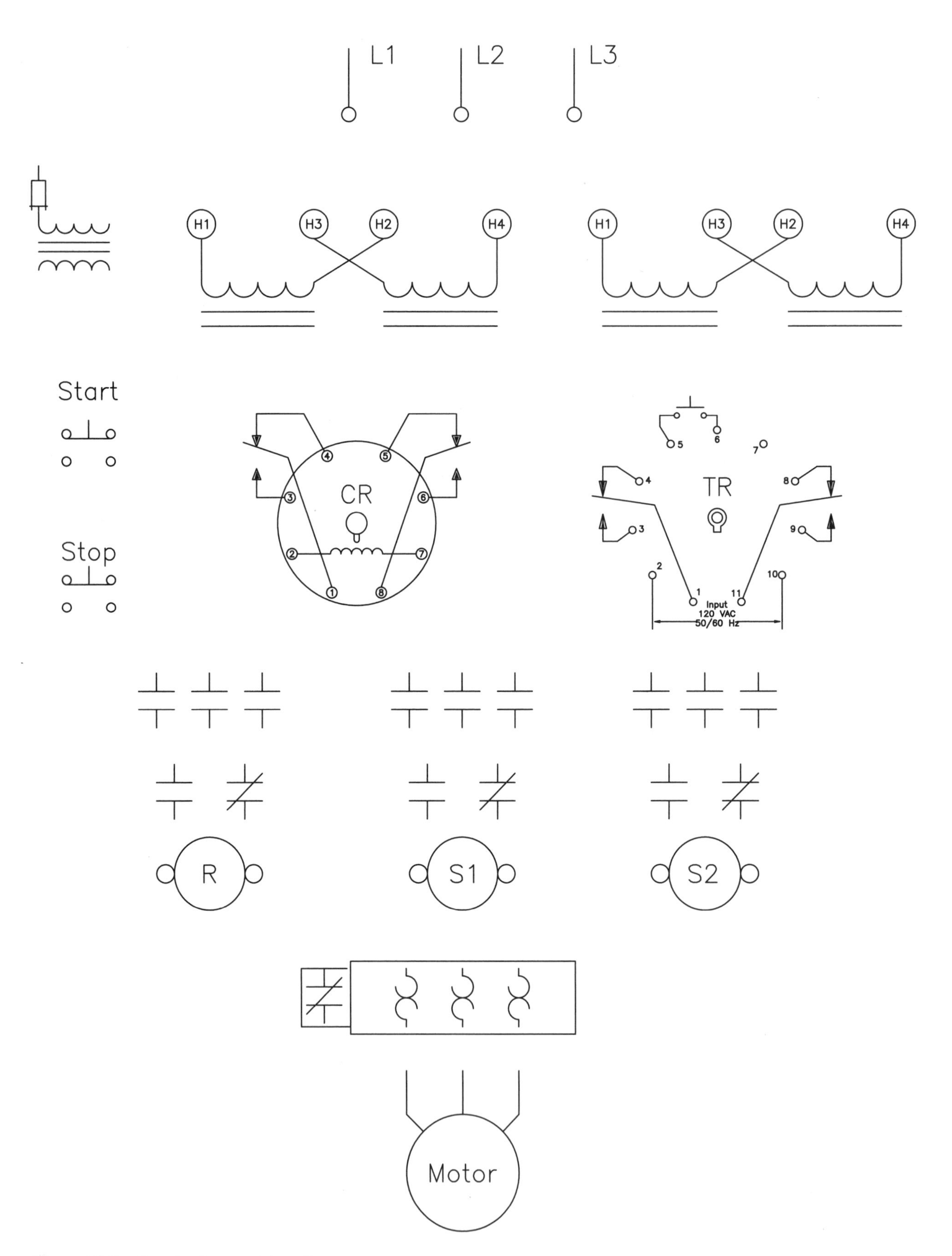

Figure 16-3 Developing a wiring diagram

Review Questions

1. How does the auto-transformer reduce the amount of starting current to a motor?

2. Is the auto-transformer used in this experiment connected as a wye, delta, or open delta?

3. What is the advantage, if any, of using an open delta connection as opposed to a closed delta or wye?

4. Assume that the line-to-line voltage in Figure 16-1 is 480 volts. Also assume that when the start button is pressed, the motor starts with 240 volts applied to the motor. When the start button is released, however, the motor stops running. Which of the following could cause this problem?
   a. S1 coil is open.
   b. CR coil is open.
   c. TR coil is open.
   d. The stop push-button is open.

5. Refer to the circuit shown in Figure 16-1. When the start button is pressed, nothing happens for a period of five seconds. After five seconds, the motor suddenly starts with full voltage connected to it. Which of the following could cause this problem?
   a. CR coil is open.
   b. TR coil is open.
   c. R coil is open.
   d. R normally closed auxiliary contact is open.

Notes

Notes

Notes

Notes

Notes

Notes

Notes

Notes

Notes

Notes

Notes

Notes

Notes

Notes

Notes

Notes